ABSTRACT OBJECTS

SYNTHESE LIBRARY

STUDIES IN EPISTEMOLOGY,
LOGIC, METHODOLOGY, AND PHILOSOPHY OF SCIENCE

VOLUME 160

EDWARD N. ZALTA
Department of Philosophy, Rice University, Houston, Texas

ABSTRACT OBJECTS

An Introduction to Axiomatic Metaphysics

D. REIDEL PUBLISHING COMPANY
A MEMBER OF THE KLUWER ACADEMIC PUBLISHERS GROUP
DORDRECHT / BOSTON / LANCASTER

Library of Congress Cataloging in Publication Data

CIP

Zalta, Edward N., 1952–
Abstract objects.

(Synthese library; v. 160)
Bibliography: p.
Includes index.
1. Metaphysics. 2. Object (Philosophy) I. Title.
BD111.Z34 1983 111 82–20549
ISBN-13: 978-94-009-6982-7 e-ISBN-13: 978-94-009-6980-3
DOI: 10.1007/978-94-009-6980-3

Published by D. Reidel Publishing Company,
P.O. Box 17, 3300 AA Dordrecht, Holland,

Sold and distributed in the U.S.A. and Canada
by Kluwer Boston Inc.,
190 Old Derby Street, Hingham, MA 02043, U.S.A.

In all other countries, sold and distributed
by Kluwer Academic Publishers Group,
P.O. Box 322, AH Dordrecht, Holland.

To my parents

"Sometimes, unexpected flashes of instruction were struck out by the fortuitous collision of happy incidents, or an involuntary concurrence of ideas, in which the philosopher to whom they happened had no other merit than that of knowing their value, and transmitting unclouded to posterity that light which had been kindled by causes out of his power."

Samuel Johnson
The Rambler
Saturday, September 7, 1751

TABLE OF CONTENTS

PREFACE xi

INTRODUCTION 1
1. Theory, Data, and Explanation 1
2. The Origins of the Theory 6

CHAPTER I. ELEMENTARY OBJECT THEORY 15
1. The Language 16
2. The Semantics 19
3. The Logic 28
4. The Proper Axioms 32
5. An Auxiliary Hypothesis 37

CHAPTER II. APPLICATIONS OF THE ELEMENTARY THEORY 40
1. Modelling Plato's Forms 41
2. Modelling the Round Square, etc. 47
3. The Problem of Existence 50
Appendix 52

CHAPTER III. THE MODAL THEORY OF ABSTRACT OBJECTS (WITH PROPOSITIONS) 59
1. The Language 59
2. The Semantics 61
3. The Logic 68
4. The Proper Axioms 73

CHAPTER IV. THE APPLICATIONS OF THE MODAL THEORY 77
1. Truth 77
2. Modelling Possible Worlds 78
3. Modelling Leibniz's Monads 84
4. Modelling Stories and Native Characters 91
5. Modality and Descriptions 99

CHAPTER V. THE TYPED THEORY OF ABSTRACT OBJECTS 107
1. The Language 109
2. The Semantics 113
3. The Logic 121
4. The Proper Axioms 124

CHAPTER VI. APPLICATIONS OF THE TYPED THEORY 126
1. Modelling Frege's Senses (I) 126
2. Modelling Frege's Senses (II) 140
3. Modelling Impossible and Fictional Relations 145
4. Modelling Mathematical Myths and Entities 147

CONCLUSION 154

APPENDICES 158
A. Modelling the Theory Itself 158
B. Modelling Notions 167

NOTES 172

BIBLIOGRAPHY 187

INDEX 190

PREFACE

In this book, I attempt to lay the axiomatic foundations of metaphysics by developing and applying a (formal) theory of abstract objects. The cornerstones include a principle which presents precise conditions under which there are abstract objects and a principle which says when apparently distinct such objects are in fact identical. The principles are constructed out of a basic set of primitive notions, which are identified at the end of the Introduction, just before the theorizing begins. The main reason for producing a theory which defines a logical space of abstract objects is that it may have a great deal of explanatory power. It is hoped that the data explained by means of the theory will be of interest to pure and applied metaphysicians, logicians and linguists, and pure and applied epistemologists.

The ideas upon which the theory is based are not essentially new. They can be traced back to Alexius Meinong and his student, Ernst Mally, the two most influential members of a school of philosophers and psychologists working in Graz in the early part of the twentieth century. They investigated psychological, abstract and non-existent objects – a realm of objects which weren't being taken seriously by Anglo-American philosophers in the Russell tradition. I first took the views of Meinong and Mally seriously in a course on metaphysics taught by Terence Parsons at the University of Massachusetts/Amherst in the Fall of 1978. Parsons had developed an axiomatic version of Meinong's naive theory of objects. The theory with which I was confronted in the penultimate draft of Parsons' book, *Nonexistent Objects*, had a profound impact on me. Parsons' work was a convincing new paradigm of philosophical investigation.

While canvassing the literature during my research for Parsons' course, I discovered, indirectly, that Mally, who had originated the nuclear/extranuclear distinction among relations (a seminal distinction adopted by both Meinong and Parsons), had had another idea which could be developed into an alternative axiomatic theory. This discovery was a result of reading both a brief description of Mally's theory in J. N. Findlay's book, *Meinong's Theory of Objects and Values* (pp. 110–112) and what appeared to be an attempt to reconstruct Mally's theory by W. Rapaport in his paper "Meinongian Theories and a Russellian Paradox". With the logical devices Parsons had used in his book, plus others learned from my colleagues or invented on my own, the alternative theory was elaborated and applied in a series of nine short working papers written between November 1978 and September 1979 (the third one was co-authored with Alan McMichael and published; the others are unpublished). Since then, the current work has been thrice drafted – once in 1979, once in 1980, and once in 1981. The first draft assimilated the nine working papers. The second was submitted as my Ph.D. dissertation at the University of Massachusetts/Amherst. The third and present draft, which preserves the essential structure of the second, was written during my stays at the University of Auckland and Rice University. This final draft is a vast improvement on its predecessors – it contains both significantly new ideas and exposition and more crystalline development of the technical material.

Chapters I, III, and V contain somewhat technical presentations of successively more powerful versions of the theory. I suggest that readers less technically inclined skip Section 2 of each of these chapters, since these contain the model-theoretic semantics which will prove useful for answering questions about the consistency of the axioms, completeness of the logic, etc. But they are not essential for understanding the primitive metaphysical and logical notions used in the statement of the axioms. Sections 1, 3, and 4 of these chapters however, contain short though valuable expositions of the language, logic, and proper axioms of the theory, respectively.

The entire project could not have been carried out without the inspiration and assistance of both teachers and colleagues. Throughout the writing of the first and second drafts, Parsons served as a sharp critic. Our conversations every couple of weeks always left me with an idea for improving my work or with an outline of a problem which had to be tackled and solved. It is to his credit that he was such a great help, despite the fact that our theories offered rival explanations to certain pieces of data. Barbara Hall Partee graciously gave her time in weekly discussions during the writing of the first draft. Her enthusiasm, encouragement, and suggestions were invaluable. My colleague, Alan McMichael, also deserves special mention. Besides teaching me the techniques of algebraic semantics, and discovering (and helping to solve) a paradox within the theory, McMichael served as my first critic. During the writing of the first two drafts, whenever I discovered a new application of the theory or got stuck on a point of logic, I frequently presented it to Alan. His criticisms and suggestions helped me to sharpen up many of the intricate details.

I would also like to thank Mark Aronszajn, Blake Barley, Max Cresswell, Cynthia Freeland, Edmund Gettier, Richard Grandy, Herbert Heidelberger, Larry Hohm, Michael Jubein, Robert Nola, Dana Scott, Krister Segerberg, and Robert Sleigh. Spirited discussions with these individuals forced me to think deeply about a variety of metaphysical and logical issues. Special thanks go to the members of the philosophy department at the University of Auckland for allowing me to lead a series of staff seminars at which the details of the theory were expounded. It was during those lectures that many ways were found to improve the second draft.

Finally, thanks goes to Nancy Scott, Helen Bedford, and Jane Addington for their dedication in typing unfriendly looking manuscripts.

December 1981 E. N. Z.
Houston

INTRODUCTION

1. THEORY, DATA, AND EXPLANATION

In this book, we shall produce a research program in metaphysics. Following Lakatos, a research program in metaphysics consists of three parts:[1] (1) a theory about what things there are, which stands at the "core" of the program; (2) a "heuristic" which grounds the principles of investigation and explanation, organizes the data to be explained, and provides a problem solving machinery for transforming recalcitrant data into confirming evidence; and (3) a "protective belt" of auxiliary hypotheses which grow out of the heuristic to guard the metaphysical theory against refutation. Our particular research program can be characterized using this framework.

The metaphysical theory which stands at the core of the program may be stated roughly as follows: in addition to existing (actual, or real) objects (like you, me, my desk, sub-atomic particles, etc.), and the properties and relations they exemplify, there are abstract entities as well. Among the abstract entities we find abstract objects (or as some might prefer to say, abstract individuals), abstract properties, and abstract relations. For the major part of this book, the theory of abstract objects is developed and applied (Chapters I–IV). At the end, the theory of abstract properties and relations is developed and applied, using the resources of a new kind of type theory (Chapters V–VI). For convenience, we refer to the overall theory as the theory of abstract objects.[2] The first principles of the theory will tell us not only the precise conditions under which there are particular abstract objects, but also the conditions under which two such objects are identical. This will provide us with a clearly defined background ontology.

The heuristic we associate with this theory revolves around two central tasks of all scientific research programs: (1) show that accepting the first principles of the theory allows us to construct explanations of interesting pieces of data, and (2) show that there is no good evidence for thinking that there must be other (kinds of) entities than those required by the theory. Such a heuristic anchors the following two methodological principles of investigation and explanation concerning the history of philosophy and the philosophy of language:

(A) History of philosophy: If earlier philosophers who postulated theoretical entities were describing anything at all, they must have been describing entities which can be found in our background ontology. So try to construe the discussions of philosophers who described theoretical entities like Forms, Monads, Possible Worlds, Senses, etc., as discussions about existing or abstract objects, properties, or relations.

(B) Philosophy of language: If the terms of a natural language denote anything at all, they must denote entities found in the background ontology. So try to show both that abstract and existing objects, properties, and relations are denoted by terms of natural language and that there are no true sentences which contain terms that denote entities other than these.

These principles leave us with philosophically interesting data to be explained.

The data consists of true sentences of natural language (and their entailments) and there are two basic kinds: the A PRIORI truths and the A POSTERIORI truths. There are two kinds of A PRIORI data. The first consists of metaphysical hypotheses, possibly developed or discussed by earlier philosophers, which we intuitively believe to be true.[3] For example, here are two hypotheses developed by earlier philosophers, followed by two hypotheses about possible worlds which we take to be true:

(i) If there are two distinct F-things, then there is a Form of F in which they both participate (Plato, *Parmenides*, 132a).

(ii) ...each simple substance (i.e., monad)...is a perpetual living mirror of the universe (Leibniz, *Monadology*, §56).

(iii) A proposition is necessarily true iff it is true in all possible worlds (Leibniz, "Necessary and Contingent Truths"?).

(iv) There is a unique actual world.

We hope to show that certain abstract objects display features resembling those of Platonic Forms, that others display features resembling Leibnizian Monads, while still others display features resembling possible worlds. By saying, for example, that certain abstract objects display features resembling those of Platonic Forms, we mean three things: (1) a definition such as the following,

$$x \text{ is a Form} =_{df} \ldots x \ldots,$$

can be given using only the primitive and defined notions of the theory; (2) features of the Forms that Plato describes are definable in the theory as well; and (3) it follows from the first principles of the theory that the abstract objects which satisfy our definition of "Form" have the features Plato says HIS Forms are supposed to have. Consequently, we shall suppose that data like (i)–(iv) have been explained if there are reasonable facsimiles of them which turn out to be consequences of the theory.

The second kind of A PRIORI data consists of sentences such as the following:

(v) The round square is round.

(vi) The set of all sets which are not members of themselves is a set of all sets which are not members of themselves.

(vii) The fountain of youth is a fountain.

Even if we have only a rough, pre-theoretical understanding of what (v)–(vii) assert, no experience is needed to decide whether they are true. (v) and (vi) are just part of the evidence we use to establish that the round square and the set of all sets which are not self-members are impossible objects. In what follows, we shall try to show that abstract objects can serve as the denotations of the descriptions in (v)–(vii). And as with all A PRIORI data, we suppose that (v)–(vii) have been explained if we can deduce them as consequences of the first principles of our theory.

The A POSTERIORI data also fall into two major groups. The first group consists of statements we ordinarily make about fictional characters, mythical figures, dream objects, and the like. Here are some examples:

(viii) Santa Claus does not exist.

(ix) Stephan Dedalus is a fictional character.

(x) In the myth, Achilles fought Hector.

(xi) Some Greeks worshipped Dionysus.

(xii) Ponce de Leon searched for the fountain of youth.

(xiii) Franz Kafka wrote about Gregor Samsa.

To construct a PRIMA FACIE case for thinking that fictional characters, mythical figures, etc., just are abstract objects, we shall focus on a formal language that we develop in Chapter III. Certain definitions tell us the conditions under which a given sentence of the formal language is true. We then translate (viii)–(xiii) into our formal language using names and descriptions in the language which denote abstract objects (as well as existing ones). Thus, an explanation of (viii)–(xiii), and others like them, consists in showing that they can be translated into sentences which preserve their intuitive truth value and their entailments.

The other group of A POSTERIORI data contains triads of sentences. These are sentences which involve verbs of propositional attitude and the "is" of identity:

(xiv) S believes that Socrates taught Plato.
(xv) S does not believe that the son of Phaenarete taught Plato.
(xvi) Socrates is the son of Phaenarete.

(xvii) S believes that Woodie is a woodchuck.
(xviii) S does not believe that Woodie is a groundhog.
(xix) Being a woodchuck just is being a groundhog.

We follow Frege in supposing that the English terms inside propositional attitude contexts do not have their ordinary denotations and that they denote their senses instead. This was Frege's explanation of why (xiv) and (xvi) do not imply the negation of (xv), and why (xvii) and (xix) do not imply the negation of (xviii). However, the senses of terms denoting objects will be construed as abstract objects and the senses of terms denoting properties (relations) will be construed as abstract properties (relations). To do this, we focus on a formal language developed in Chapter V. (xiv)–(xix) are translated into our language, using names and descriptions of the language which denote abstract entities that serve as the senses of English terms. Thus, our Fregean explanation of the consistency of each triad lies in showing that the sentences which translate the members of a given triad are consistent.

These, then, will be the kinds of data and explanation which shall occupy our attention. In the course of solving problems and broadening the scope of the theory, we will adopt auxiliary hypotheses in both metaphysics and the philosophy of language. But we can not describe these even in a rough way in advance of the rigorous presentation of the theory.

The research program we have just outlined has been designed to compete with the current programs in metaphysics. No attempt will be made to provide a list of these alternatives, but it would serve well to mention just a few. The most influential one has developed around a theory that many philosophers have attributed to Russell, namely, that existing objects, and the properties and relations they exemplify, are the only things there are.[4] We shall sometimes talk as if the established "Russellian view" on certain issues is the only one to take seriously, partly because this view has been so widely regarded as true, but partly because other influential research programs, which initially appear to differ from the Russellian view, are nevertheless closely allied to it. (For example, one research program has developed around the metaphysical theory that individuals, sets, and possible worlds are the only things there are.[5] While this ontology differs from the classic Russellian ontology, philosophers in this tradition reconstruct properties and relations out of functions from possible worlds to sets of sequences of individuals and thereby provide a link with Russellian metaphysics.) In addition to the Russellian program and its modern counterparts, there is also the physicalist program, which guards the theory that the entities described by the correct physics are the only things there are (the less radical physicalists commit themselves to sets as well).[6] And finally, we should note that there are other programs besides ours which have been motivated by Meinong's views on what there is.[7]

Among a host of competing research programs, Lakatos would distinguish those that are "progressive" from those that are "degenerating". Basically, the distinction is that in a progressive program, the addition of auxiliary hypotheses does not just accommodate known facts and anomalies, but also leads to the discovery of hitherto unknown, novel facts. Research programs may go through phases of progression, degeneration, and then progression again, and it is hard to tell whether a current phase of degeneration will be a permanent one. So the fact that a particular program is presently in a phase of degeneration is not sufficient reason for switching to an alternative program. However, it is good reason for careful scrutiny of the successes of alternative programs.

In what follows, we shall not argue that all other programs are in permanent phases of degeneration. Nor shall we argue that the data we have presented is not, and cannot, be assimilated by the other programs. Enough AD HOC auxiliary hypotheses can be appended to any theory to enable it to handle the data. Instead, we will try to establish one thesis, namely, that the research program in metaphysics developed here is progressive one (i.e., our theory helps us to explain our data and, together with auxiliary hypotheses, predicts hitherto unknown, novel facts). Then, any philosopher, linguist, or cognitive scientist who agrees that the data we have chosen are important, interesting, and currently lack natural explanations, should either consider our program as a viable alternative or be prompted to find progressive explanations from within current programs.

The above description should give the reader a rough idea of the nature of our project. But in the course of setting up our theory, many philosophical issues will be confronted. For example, we will end up developing a full-fledged theory of relations (where properties and propositions turn out to be one-place and zero-place relations, respectively). This includes a definition which tells us when any two relations (properties, propositions) are in fact the same. Semantics for our formal languages are developed in which we may consistently suppose that logically equivalent relations are distinct. The resulting metaphysical system should be attractive not only because it might handle important kinds of data which seem problematic for current traditions, but also because it exhibits many interesting and philosophically satisfying qualities in its own right.

2. THE ORIGINS OF THE THEORY

The theory we will develop has its origins directly in the naive theory of nonexistent objects which Meinong and Mally investigated at the turn of the century. A very simple statement of the theory upon which Meinong seemed to be relying in his early work is the following, which we call Naive Object Theory:

(NOT) For every describable set of properties, there is an object which exemplifies just the members of the set.

For example, there is an object which exemplifies just the properties in the set which contains only the two properties roundness and squareness. If we round out (NOT) by assuming "Leibniz' Law", namely that two objects are identical if and only if they exemplify exactly the same properties, then such an object would be unique. Maybe Meinong's idea was that this object is what is being talked about or what is denoted when the definite description "the round square" is used. (NOT) also guarantees an object which exemplifies just the properties in the set of properties-attributed-to-Zeus-in-the-myth. Maybe this object is denoted by the name "Zeus". It is important here for the reader to try to anticipate how (NOT) might be used to construct explanations of some of the other data we presented in Section 1.

(NOT) may be stated precisely in a second order predicate calculus without directly invoking the notion of a set or the set membership relation. Such a statement would be preferable to philosophers like myself who want to initially remain neutral on the ontological status of mathematical entities like sets, but who are willing to commit themselves to objects (in general) and relations. A second order predicate calculus is an interpreted, deductive system built upon a second order language in which the following three metaphysically primitive notions are embedded:

> ***object***: $x, y, z, \ldots$
> ***n-place relation***: $F^n, G^n, H^n, \ldots$
> (where ***properties*** $=_{df}$ 1-place relations: $F^1, G^1, H^1, \ldots$)
> $x_1, \ldots, x_n$ ***exemplify*** F^n: $F^n x_1 \ldots x_n$.

The symbolic representations of the notion of exemplification ("$F^n x_1 \ldots x_n$") serve as the atomic statements of the formal language. The language also utilizes three primitive logical notions:

> ***It is not the case that*** ϕ: $\sim \phi$.
> ***If*** ϕ, ***then*** ψ: $\phi \rightarrow \psi$.
> ***Every*** x(***every*** F^n) is such that ϕ: $(\forall x)\phi$, $(\forall F^n)\phi$.

We frequently abbreviate $(\forall x)\phi$ as $(x)\phi$ and $(\forall F^n)\phi$ as $(F^n)\phi$. The other logical notions of our basic predicate calculus such as ***both*** ϕ ***and*** ψ("ϕ & ψ"), ϕ ***or*** ψ("$\phi \vee \psi$"), ϕ ***if-and-only-if*** ψ("$\phi \equiv \psi$"), and ***some*** x(***some*** F^n) is ***such that*** ϕ ("$(\exists x)\phi$), $(\exists F^n)\phi$"), can all be defined in the standard way. The simple and complex statements which can be constructed out of these primitive and defined notions reveal the expressive power of the system.

To see how to express (NOT) precisely in this system it is important to first look at a general method for describing sets of properties which employs our second order language. Then we can indicate how to represent (NOT) without mentioning sets. Consider the following open formula "Socrates exemplifies F^1". If we let "s" denote Socrates, then we can represent this condition on properties in our language as "$F^1 s$". Now we can form the following description of a set: the set of all properties F^1 such that Socrates exemplifies F^1, i.e., $\{F^1 | F^1 s\}$. This describes the set of properties which satisfy (in Tarski's sense) the open condition "$F^1 s$." The set contains properties like being a philosopher, being Greek, being snub-nosed, etc. Here's another example, where "p" denotes Plato. Take the open condition "both Socrates exemplifies F^1 and Plato exemplifies

F^1" ("F^1s & F^1p") and form the set abstract: the set of all properties F^1 such that both Socrates exemplifies F^1 and Plato exemplifies F^1, i.e., $\{F^1|F^1s \,\&\, F^1p\}$. The set described here contains such properties as being a philosopher and being Greek as well, but it would not contain the property of being snub-nosed, since Plato did not exemplify that property (let us suppose). Consider another example using the second order language with identity. We can form the open condition "either F^1 is identical to the property of being round or F^1 is identical to the property of being square" ("$F^1 = R^1 \vee F^1 = S^1$"). Then form the description: the set of all properties F^1 such that F^1 is identical to roundness or F^1 is identical to squareness, i.e., $\{F^1|F^1 = R^1 \vee F^1 = S^1\}$. This describes a set which contains just the two properties satisfying the open condition, namely roundness and squareness. We can even consider formulas without a free property variable to express vacuous conditions on properties. For example, if "P" denotes being a philosopher, then every property is a member of $\{F^1|Ps\}$. In this manner, any condition on properties expressible in our language can be used to describe sets of properties.

Now (NOT) asserts that for each such set of properties, there is an object which exemplifies all and only the properties in the set. This assertion can be captured in our second order language without mentioning sets by using an axiom schema. An axiom schema is basically a rule which says that every sentence of a certain form shall be considered to be an axiom. Let ϕ be any condition on properties expressible in our language (possibly with identity), as in the above examples. Then any instance of the following sentence schema is to be an axiom:

(NOT′) $(\exists x)(F^1)(F^1x \equiv \phi)$, where ϕ has no free x's.

Here are four instances of (NOT′) which guarantee that there are objects which correspond to the sets of properties described in the above examples:[8]

(a) $(\exists x)(F^1)(F^1x \equiv F^1s)$

(b) $(\exists x)(F^1)(F^1x \equiv F^1s \,\&\, F^1p)$

(c) $(\exists x)(F^1)(F^1x \equiv F^1 = R^1 \vee F^1 = S^1)$

(d) $(\exists x)(F^1)(F^1x \equiv Ps)$.

A complete theory of objects may be obtained by adding Leibniz's Law to the infinite set of axioms generated by (NOT′):

(LL) $x = y \equiv (F^1)(F^1x \equiv F^1y)$.

(LL) ensures that the object yielded by an arbitrary instance of (NOT′) is unique, since there couldn't be distinct objects which exemplify exactly the properties satisfying the given condition ϕ.

Although we shall not specifically attribute the theory which results to Meinong, it does appear to be the natural way to formalize the principles upon which he seemed to be relying. Unfortunately, there are lots of things wrong with the theory, and philosophers since Russell, who have worked with informal versions of the theory, have been quick to recognize this. For one thing, (a) yields an object exemplifying just the properties Socrates exemplifies. By (LL), any such object just is Socrates. So the A PRIORI metaphysics rules that Socrates has being. Yet the being of Socrates seems to be a contingent matter. Secondly, (c) is incompatible with the natural assumption that whatever exemplifies roundness fails to exemplify squareness. (c) also implies that Russell never thought about the round square (on the natural assumption that the property of being thought about by Russell is distinct from both the property of being round and being square). This is dubious, at best.

But there are much more serious difficulties with the theory. It implies falsehoods and is incompatible with a very important principle yielding complex relations. Consider first the following instance of (NOT′), noted by Russell, where "*E*!" stands for existence, "*G*" stands for goldenness, and "*M*" stands for mountainhood:

$$(\exists x)(F)(Fx \equiv F = E! \vee F = G \vee F = M).$$

This implies the falsehood that there is a golden mountain which exists. But the most serious problem with (NOT′) is that it is inconsistent with the following abstraction schema for relations:

$$(\exists F^n)(x_1)\ldots(x_n)(F^n x_1 \ldots x_n \equiv \phi), \text{ where } \phi \text{ has no free } F^n\text{'s.}$$

Here are two typical instances of this schema:

$$(\exists F)(x)(Fx \equiv {\sim} Gx)$$
$$(\exists F)(x)(Fx \equiv Gx \;\&\; Hx).$$

The first guarantees that any given property G will have a negation, while the second guarantees that any two given properties G and H will have a conjunction. There are many other kinds of complex properties yielded by this schema as well. But consider, in particular, the following:

$$(\exists F)(x)(Fx \equiv Rx \;\&\; {\sim} Rx).$$

This says that there is a property objects exemplify iff they exemplify redness and it is not the case that they exemplify redness. Call an arbitrary such property K. The assumption that something exemplifies K produces an immediate contradiction. But (NOT′) ensures just that:

$$(\exists x)(F)(Fx \equiv F = K).$$

So if we want to keep our abstraction schema for relations as it is stated, we have to give up (NOT′) (there are other interesting ways to produce contradictions from (NOT′) and the relations schema, but we shall not discuss them here).

One suggestion by Mally to refine (NOT) was to distinguish two types of properties – nuclear and extranuclear.[9] The nuclear properties an object has are its "ordinary" properties and are more central to its nature and identity than its extranuclear properties. Terence Parsons follows up on this suggestion.[10] He adds this distinction as one new primitive to a standard second order predicate calculus. He develops the theory and logic associated with nuclear and extranuclear relations in general. He restricts the property quantifier in (NOT′) so that it ranges just over nuclear properties. He also restricts the relation quantifier in the abstraction schema for complex relations so that it ranges just over extranuclear relations. His theory is based on the following three principles, where "F^n" ranges over extranuclear n-place relations and "f^n" ranges over nuclear n-place relations:[11]

(I) For every condition on nuclear properties, there is an object which exemplifies just the properties satisfying the condition

$$(\exists x)(f^1)(f^1 x \equiv \phi), \text{ where } \phi \text{ has no free } x\text{'s}.$$

(II) Two objects are identical iff they exemplify the same nuclear properties

$$x = y \equiv (f^1)(f^1 x \equiv f^1 y).$$

(III) For every extranuclear relation, there is a nuclear relation which is coextensive with it on the existing objects

$$(F^n)(\exists f^n)(x_1)\ldots(x_n)(E!x_1 \ \& \ldots \& \ E!x_n \rightarrow (f^n x_1 \ldots x_n \equiv F^n x_1 \ldots x_n)).$$

So on Parsons theory, there are two kinds of relations and one kind of object. Principle III yields a nuclear, "watered down" version of each extranuclear relation, but it is a consequence of the theory that distinct extranuclear relations sometimes have the same nuclear watered down version.

Consequently, Parsons avoids generating the above oddities, falsehoods, and inconsistencies. Although the theory asserts that there is an object which exemplifies just the nuclear properties Socrates exemplifies and that this object just is Socrates, it is not a consequence of the theory that Socrates exemplifies extranuclear existence. The object which exemplifies just nuclear existence, nuclear goldenness, and nuclear mountainhood, ("the existent golden mountain") provably does not exemplify extranuclear existence. (Nuclear existence is just the watered down version of extranuclear existence.) The object which exemplifies just nuclear roundness and nuclear squareness could, and in fact did, exemplify the extranuclear property of being thought about by Russell (intentional properties are classic extranuclear properties). This object doesn't violate the principle that everything exemplifying roundness fails to exemplify squareness because the principle is false when the quantifier "everything" is allowed to range over nonexistent objects as well as existing objects (of course, all existing objects exemplifying roundness fail to exemplify squareness). Also, the only way to produce the property *K* described above is to use the abstraction principle for complex extranuclear relations (i.e., use the principle we used to produce *K* in the first place, except now, the variable "F^n" ranges just over extranuclear relations). So *K* is an extranuclear property and the theory does not imply that there is an object which exemplifies being-red-and-not-being-red. And with these obstacles out of the way, Parsons finds interesting applications for his theory. In particular, he models fictional characters (and the like), Leibnizian Monads, and suggests how to model Plato's Forms.[12] These models served as prototypes for the models we have constructed in our alternative object theory.

Our theory of abstract objects is based on an entirely different suggestion of Mally's, however. He distinguished two relationships which relate objects to their properties. On Mally's view, properties can DETERMINE objects which do not in turn SATISFY the properties.[13] For example, the properties roundness and squareness can determine an abstract object which satisfies neither roundness nor squareness. The properties of existence, goldenness, and mountainhood can determine an abstract object which does not satisfy any of these properties. The properties which determine an abstract object are central to its identity. For a recent attempt to reconstruct Mally's theory, see W. Rapaport's discussion in "Meinongian Theories and a Russellian Paradox".[14]

In what follows, we construct languages capable of representing the distinction between satisfying and being determined by a property. However, we shall employ different terminology. We shall say that an object ***exemplifies*** a property instead of satisfying it. We shall say that an object ***encodes*** a property instead of saying that the object is determined by the property. The distinction between exemplifying and encoding a property is a primitive one and will be represented by a distinction in atomic formulas of the languages we construct. All the primitive notions that we shall need in order to state the first principles of the theory are listed and followed by their symbolic representations:

Primitive Metaphysical Notions

object: $x, y, z, \ldots$
***n*-place relation:** $F^n, G^n, H^n, \ldots$
$x_1, \ldots, x_n$ **exemplify** F^n: $F^n x_1 \ldots x_n$
x **encodes** F^1: xF^1.

Primitive Logical Notions

it is not the case that ϕ: $\sim \phi$
if ϕ, **then** ψ: $\phi \rightarrow \psi$.
every x (**every** F^n) **is such that** ϕ: $(\forall x)\phi$, $(\forall F^n)\phi$.

Primitive Theoretical Relations

existence: $E!$.

Using these basic notions, we define a ***property*** to be a one-place relation and say that x ***is abstract*** ("$A!x$") iff x fails to exemplify existence.[15] We also say that two objects x and y are ***identical***$_E$ ("$x =_E y$") iff x and y both exemplify existence and exemplify the same properties (for reasons which we cannot go into here, in Chapter I we will take $=_E$ as a primitive two-place theoretical relation and cast the preceding definition as a proper axiom).

Now if we have understood Mally's insight correctly, the main principle of the theory must assert that for every condition on properties, there is an abstract object which is determined by just the properties meeting the condition. Using our new terminology, this can be captured by the following principle:

(I) For every expressible condition on properties, there is an abstract object which encodes just the properties meeting the condition:

$(\exists x)(A!x \;\&\; (F^1)(xF^1 \equiv \phi))$, where ϕ has no free x's.

The two other principles which serve as the cornerstones of the theory of abstract objects are:[16]

(II) Two objects are identical iff they are identical$_E$ or they are both abstract and encode the same properties

$$x = y \equiv x =_E y \vee (A!x \;\&\; A!y \;\&\; (F^1)(xF^1 \equiv yF^1)).$$

(III) Two properties are identical iff they are encoded by the same objects

$$F^1 = G^1 \equiv (x)(xF^1 \equiv xG^1).$$

Principle (I) gives us "being" conditions for abstract objects. Principle (II) gives us identity conditions for all objects. And Principle (III) yields identity conditions for properties (in Chapter I, we will generalize on this definition to obtain an identity principle for all relations). On our theory, in contrast to Parsons, there is just one kind of relation and we will avail ourselves of the abstraction schema for complex relations described above without restricting the variable "F^n" in any way (though, in order to avoid paradoxes, we shall not allow any new relations to be constructed using encoding formulas – only the relations constructible in the standard second order calculus will be found).

With these principles, we will find an abstract object which encodes just the properties Socrates exemplifies ($(\exists x)(A!x \;\&\; (F)(xF \equiv Fs))$). But, clearly, this object is not identical with Socrates. We also find an abstract object which encodes just roundness and squareness ($(\exists x)(A!x \;\&\; (F)(xF \equiv F = R \vee F = S))$). But our principles do not imply that this object exemplifies either of these properties. They are compatible with the claim that everything whatsoever which exemplifies roundness fails to exemplify squareness. We also find an abstract object which encodes just existence, goldenness, and mountainhood ($(\exists x)(A!x \;\&\; (F)(xF \equiv F = E! \vee F = G \vee F = M))$). Although the theory presupposes that this object fails to exemplify existence, this is compatible with the contingent fact that no existing object exemplifies all the properties this abstract object encodes (which is how we will read the ENGLISH nonexistence claim). Finally, the abstraction principle for complex relations will generate the property K in exactly the manner described above and the theory will guarantee that there is an object which encodes K ($(\exists x)(A!x \;\&\; (F)(xF \equiv F = K))$). But it is provable that this object does not exemplify this property. Indeed, it's provable that no object does, and our principles are compatible with this result.

In Chapter I, we will couch principles (I), (II), and (III) as a proper axiom schema and two definitions, respectively. The abstraction schema for complex relations will be a logical theorem schema. We also axiomatize the other logical and non-logical principles which round out the theory. This should make the details of an ontology rich with abstract objects sharp and accurate. Once our background ontology is set, we will go on to apply the theory in Chapter II.

CHAPTER I

ELEMENTARY OBJECT THEORY

The full presentation of the elementary theory of abstract objects shall occupy the first four sections of this chapter. In each of these sections, we concentrate on the following major groups of definitions:

1. The Language.
2. The Semantics.
3. The Logic.
4. The Proper Axioms.

The proper axioms are stated in the language. Since the semantics contains a definition which tells us the conditions under which an arbitrary formula of the language is true, we will know what is being asserted by our proper axioms. The logic we associate with the language allows us to prove the consequences of the proper axioms.

In the course of the definitions which follow, we frequently provide examples and make extended remarks to explain and motivate unusual features. In the remarks, we frequently define (with the help of **boldface** and lists) certain syntactic or semantic concepts which will help us to single out classes of expressions or entities which have certain properties. We use quotation marks to mention expressions of the language. We generally omit these standard devices (quotation marks, corner marks) for mentioning and describing pieces of language when the intent is clear. We use quotation marks inside parentheses ("…"), to give readings and/or abbreviations of formulas. All definitions of the object language appear with the label " $\boldsymbol{D}_n$."

With the exception of "λ" and "ι", we use lower case Greek letters to range over expressions of the formal language. In particular, we use:

κ's to range over names (i.e., constants)
ϕ, ψ, χ, θ to range over formulas
o's to range over object terms
ρ^n's to range over relation terms
α, β, γ to range over all variables
τ's to range over all terms

ν's to range over object variables
π^n's to range over relation variables
μ, ξ, ζ to range over λ-expressions.

Finally, we note that in most of Chapter I, we shall not give the intuitive readings in natural language of the formulas and complex terms of the object language. That is because our aim is to focus on the expressive capacity of a formal language, without prejudice as to how English sentences and terms are to be translated into the language. However, it will be useful to provide some examples in natural language, since this will help the reader to picture what the language and theory can, and ultimately will, say.

1. THE LANGUAGE

We shall utilize a slightly modified second order language. The ONLY modification is that new atomic formulas have been added – they express the fact that an object encodes a property. These new atomic formulas are called "encoding formulas", and whereas the ordinary "exemplification formulas" which we shall still have around have n object terms to the right of an n-place relation term, encoding formulas have a single object term to the left of a ONE-place relation term. These atomic encoding formulas can combine with other formulas to make molecular and quantified formulas. The complex formulas which result may be constructed solely out of atomic exemplification subformulas, solely out of atomic encoding subformulas, or may be of mixed construction. Many of the interesting definitions, axioms, and theorems are mixed formulas.

Our language will also have one kind of complex term – the λ-expressions. These terms shall denote relations, and they involve the primitive logical notion "being such that", which logicians represent with the λ. However, only complex formulas constructed solely out of atomic exemplification formulas can combine with the λ to form relation terms (for reasons to be explained shortly).

The language which results has much more expressive capacity than the standard second order language with complex terms. The definitions which precisely describe the language may be subdivided as follows:

A. Primitive Symbols.
B. Formulas and Terms.

A. PRIMITIVE SYMBOLS

We have two kinds of **primitive object terms: names** and **variables.** Officially we use the subscripted letters $a_1, a_2, a_3, \ldots$ as primitive object names, but unofficially, we use $a, b, c, \ldots$ for convenience. Officially, we use the subscripted letters $x_1, x_2, x_3, \ldots$ as primitive object variables, but unofficially we use $x, y, z, \ldots$. There are also two kinds of **primitive relation terms: names** and **variables.** Officially, we use the superscripted and subscripted letters $P_1^n, P_2^n, \ldots, n \geq 1$, as primitive relation names (unofficially: $P^n, Q^n, \ldots$) and $F_1^n, F_2^n, \ldots, n \geq 1$, as primitive relation variables (unofficially: $F^n, G^n, \ldots$). $E!$ is a distinguished one-place relation name; $=_E$ is a distinguished two-place relation name.

In addition we use two connectives, $\sim$, and $\rightarrow$; a quantifier: $\forall$; a lambda: λ; and we avail ourselves of parentheses and brackets to disambiguate.

B. FORMULAS AND TERMS.

We present a simultaneous inductive definition of (**propositional**) **formula, object term**, and **n-place relation term**.
The definition contains six clauses:

(1) All primitive object terms are object terms and all primitive n-place relation terms are n-place relation terms.

(2) Atomic exemplification: If ρ^n is any n-place relation term, and $o_1, \ldots, o_n$ are any object terms, $\rho^n o_1 \ldots o_n$ is a (propositional) formula (read: "$o_1, \ldots, o_n$ exemplify relation ρ^n").

(3) Atomic encoding: If ρ^1 is any one-place relation term and o is any object term, $o\rho^1$ is a formula (read: "o encodes property ρ^1").

(4) Molecular: If ϕ and ψ are any (propositional) formulas, then $(\sim \phi)$ and $(\phi \rightarrow \psi)$ are (propositional) formulas.

(5) Quantified: If ϕ is any (propositional) formula, and α is any (object) variable, then $(\forall \alpha)\phi$ is a (propositional) formula.

(6) Complex n-place relation terms: If ϕ is any propositional formula with n-free object variables $v_1, \ldots, v_n$, then $[\lambda v_1 \ldots v_n \phi]$ is an n-place relation term.

We rewrite atomic exemplification formulas of the form $=_E o_1 o_2$ as $o_1 =_E o_2$. We drop parentheses to facilitate reading complex formulas whenever there is little potential for ambiguity. We utilize the standard abbreviations: $(\phi \,\&\, \psi)$, $(\phi \equiv \psi)$, $(\phi \vee \psi)$, and $(\exists\alpha)\phi$. And we define:

D_1 *x is abstract* ("$A!x$") $=_{df} [\lambda y \sim E!y]x$.

Here then are some examples of formulas: P^3axb ("a, x, and b exemplify relation P^3"); aG ("a encodes property G"); $\sim(\exists x)(xQ \,\&\, Qx)$ ("no object both encodes and exemplifies Q"); $(x)(E!x \rightarrow \sim(\exists F)xF)$ ("every object which exemplifies existence fails to encode any properties"); and $(\exists x)(A!x \,\&\, (F)(xF \equiv Fa))$ ("some abstract object encodes exactly the properties a exemplifies").

By inserting all the parenthetical remarks when reading the above definition, we obtain a definition of **propositional formula**. In effect, a formula ϕ is propositional iff ϕ has no encoding subformulas and ϕ has no subformulas with quantifiers binding relation variables.[1] Only propositional formulas may occur in λ-expressions. λ-expressions allow us to name complex relations.[2] We read $[\lambda v_1 \ldots v_n \phi]$ as "being objects $v_1, \ldots, v_n$ such that $\phi(v_1, \ldots, v_n)$", or as "being a first thing, second thing,..., and n^{th}-thing such that ϕ". For example: $[\lambda x \sim Rx]$ ("being an object x such that x fails to exemplify R"); $[\lambda x\, Px \,\&\, Qx]$ ("being an object x such that x exemplifies both P and Q"); $[\lambda x\, x =_E b]$ ("being identical$_E$ with b"); $[\lambda xy\, Px \,\&\, Syx]$ ("being objects x and y such that x exemplifies P and y bears S to x"); $[\lambda x (\exists y) Fxy]$ ("being an x such that x bears F to something"); $[\lambda xyz\, Gzx \,\&\, E!y]$ ("being a first, second, and third thing such that the third bears G to the first and the second exists").[3]

Since arbitrary formulas ϕ cannot appear after λ's, the following expressions are ill-formed: $[\lambda x\, xP]$, $[\lambda y\, yP \,\&\, Py]$, $[\lambda x (\exists F)Fx]$, and $[\lambda x (\exists F)(xF \,\&\, \sim Fx)]$. The first two are ill-formed because the formula after the λ has an encoding subformula; the third because the formula contains a quantifier binding a relation variable ("relation quantifier"); the fourth fails both "restrictions" on propositional formulas. The "no encoding subformulas" restriction is essential. It serves to prevent paradoxes in the presence of the proper axioms. For a detailed discussion of the paradoxes this move eliminates, see Appendix A, part A (especially the discussion concerning "Clark's Paradox"). The "no relation quantifiers" restriction is not essential. However, it allows us to effect a huge simplification of the semantics. Since we shall not critically need to use λ-expressions with relation quantifiers in the applications of the theory, we choose not to complicate the semantics any further than necessary. We shall show how to bypass this restriction once we move to the typed theory of abstract objects (Chapter V). The semantics for the language which couches the typed theory more easily assimilates the interpretation of λ-expressions with "higher-order" quantifiers. The net result of these restrictions is that no relation denoting expression not already found in the standard second order language can be constructed. Intuitively, this means we will be working with familiar sorts of complex properties and relations.

These λ-expressions widen the possibilities for atomic and complex formulas: $[\lambda xy\, Px \,\&\, Qy]ab$ ("a and b exemplify being two objects x and y such that x emplifies P and y exemplifies Q"); $x[\lambda y \sim Ry]$ ("x encodes failing-to-exemplify-R"); $(\exists x)(A!x \,\&\, (F)(xF \equiv (\exists G^2)((Gab \,\&\, F = [\lambda y\, Gyb]) \vee (Gba \,\&\, F = [\lambda y\, Gby]))))$ ("some abstract object encodes just the relational properties a exemplifies with respect to b").

Finally, we say that τ is a ***term*** iff τ is an object term or there is an n such that τ is an n-place relation term.

In the definitions which follow in Sections 2, 3, and 4, it shall be useful to have precise definitions for certain syntactic concepts which up until now, we have used on an intuitive basis:[4] all and only formulas and terms are **well-formed expressions**. An occurrence of a variable α in a well-formed expression is **bound (free)** iff it lies (does not lie) within a formula of the form $(\forall\alpha)\phi$ or a term of the form $[\lambda v_1 \ldots \alpha \ldots v_n \phi]$ within the expression. A variable is **free (bound)** in an expression iff it does (does not) have a free occurrence in that expression. A **sentence** is a formula having no free variables.

Furthermore, a term τ is said to be **substitutable for** a variable α **in** a formula ϕ iff for every variable β free in τ, no free occurrence of α in ϕ occurs either in a subformula of the form $(\forall\beta)\psi$ in ϕ or in a term $[\lambda v_1 \ldots \beta \ldots v_n \psi]$ in ϕ. Intuitively, if τ is substitutable for α in ϕ, no free variable β in τ gets "captured" when τ is substituted for α, by a quantifier or λ in ϕ which binds β. We write $\phi(\alpha_1, \ldots, \alpha_n)$ to designate a formula which may or may not have $\alpha_1, \ldots, \alpha_n$ occuring free. Finally, we write $\phi^{\tau_1, \ldots, \tau_n}_{\alpha_1, \ldots, \alpha_n}$ to designate the formula which results when, for each i, $1 \leq i \leq n$, τ_i is substituted for each free occurrence of α_i in ϕ.

2. THE SEMANTICS

The definitions which help to determine the conditions under which the formulas of the language are true may by grouped as follows:

A. Interpretations.
B. Assignments and Denotations.

C. Satisfaction.
D. Truth under an interpretation.

In the definitions which follow, we use script letters as names and variables for sets, entities, and functions which are all peculiarly associated with the semantics.

A. INTERPRETATIONS[5]

An interpretation, $\mathscr{I}$, of our language is any 6-tuple $\langle \mathscr{D}, \mathscr{R}, ext_{\mathscr{R}}, \mathscr{L}, ext_{\mathscr{A}}, \mathscr{F} \rangle$ which meets the conditions described in this subsection. The first two members, $\mathscr{D}$ and $\mathscr{R}$, must be non-empty classes – they provide entities for the primitive and complex names of the language to denote and they serve as the domains of quantification. $\mathscr{D}$ is called the **domain of objects**, and we use o's as metalinguistic variables ranging over members of this domain. $\mathscr{R}$ is called the **domain of relations** and it is the union of a sequence of non-empty classes $\mathscr{R}_1, \mathscr{R}_2, \mathscr{R}_3, \ldots$; i.e., $\mathscr{R} = \bigcup_{n \geq 1} \mathscr{R}_n$. Each $\mathscr{R}_n$ is called the class of n-place relations. We use "r^n" as a metalinguistic variable ranging over the elements of $\mathscr{R}_n$. $\mathscr{R}$ must be closed under all of the logical functions specified in the fourth member of the interpretation ($\mathscr{L}$).

Intuitively, the third, fourth, and fifth members of any interpretation are functions (or classes of functions) which impose a certain structure on the elements of $\mathscr{D}$ and $\mathscr{R}$. We suppose that for each n-place relation in $\mathscr{R}_n$, there is a set of n-tuples of elements drawn from $\mathscr{D}$ which serves as the **exemplification extension** ("extension$_{\mathscr{R}}$") of the relation. Each n-tuple of the set represents an ordered group of objects which exemplify (bear, stand in) the relation. The third member of an interpretation is therefore a function, $ext_{\mathscr{R}}$, which maps each $r \in \mathscr{R}_n$ into $\mathfrak{P}(\mathscr{D}^n)$ ("the power set of $\mathscr{D}^n$"), i.e., $ext_{\mathscr{R}}: \mathscr{R}_n \rightarrow \mathfrak{P}(\mathscr{D}^n)$. We call $ext_{\mathscr{R}}(r^n)$ the exemplification extension of r^n.

The fourth member of any interpretation, $\mathscr{L}$, is a class of logical functions which operate on the members of $\mathscr{R}_n$ and $\mathscr{D}$ to produce the complex relations which serve as the denotations for the λ-expressions. Each complex relation receives an exemplification extension which must mesh, in a natural way, with the extensions$_{\mathscr{R}}$ of the simpler relations it may have as parts.

There are six elements in $\mathscr{L}$ – the first four are each families of indexed logical functions:[6] $\mathscr{PLUG}_i$ ("i-plug"), $\mathscr{UNIV}_i$ ("i-universalization"), $\mathscr{CONV}_{i,j}$ ("i, j-conversion"), and $\mathscr{REFL}_{i,j}$ ("i, j-reflection"), where i, j are elements of the set of natural numbers. The other two members of $\mathscr{L}$ are particular functions $\mathscr{COND}$ ("conditionalization") and $\mathscr{NEG}$ ("negation"). These six elements of $\mathscr{L}$ work as follows:

(a) $\mathscr{PLUG}_1$ maps $(\mathscr{R}_2 \cup \mathscr{R}_3 \cup \ldots) \times \mathscr{D}$ into $(\mathscr{R}_1 \cup \mathscr{R}_2 \cup \ldots)$. $\mathscr{PLUG}_j$, for each $j > 1$, maps $(\mathscr{R}_j \cup \mathscr{R}_{j+1} \cup \ldots) \times \mathscr{D}$ into $(\mathscr{R}_{j-1} \cup \mathscr{R}_j \cup \ldots)$.
$\mathscr{PLUG}_i$ is subject to the following condition:
$ext_{\mathscr{R}}(\mathscr{PLUG}_i(\imath^n, o)) = \{\langle o_1, \ldots, o_{i-1}, o_{i+1}, \ldots, o_n \rangle \mid \langle o_1, \ldots, o_{i-1}, o, o_{i+1}, \ldots, o_n \rangle \in ext_{\mathscr{R}}(\imath^n)\}$.

The condition on $\mathscr{PLUG}_i$ basically says that the extension$_{\mathscr{R}}$ of the new relation $\mathscr{PLUG}_i(\imath^n, o)$ ("the ith-plugging of $\imath^n$ by o") includes just those $n-1$ tuples which result by deleting the object o from the ith place of every n-tuple in the extension$_{\mathscr{R}}$ of the original relation $\imath^n$ which has o in its ith place. This ensures, for example, that an object o_1 which falls in the extension$_{\mathscr{R}}$ of the property $\mathscr{PLUG}_2(\imath^2, o_5)$ is such that $\langle o_1, o_5 \rangle$ is in the extension$_{\mathscr{R}}$ of $\imath^2$.

(b) $\mathscr{UNIV}_1$ maps $(\mathscr{R}_2 \cup \mathscr{R}_3 \cup \ldots)$ into $(\mathscr{R}_1 \cup \mathscr{R}_2 \cup \ldots)$. $\mathscr{UNIV}_j$, for each $j > 1$, maps $(\mathscr{R}_j \cup \mathscr{R}_{j+1} \cup \ldots)$ into $(\mathscr{R}_{j-1} \cup \mathscr{R}_j \cup \ldots)$.
$\mathscr{UNIV}_i$ is subject to the condition: $ext_{\mathscr{R}}(\mathscr{UNIV}_i(\imath^n)) = \{\langle o_1, \ldots, o_{i-1}, o_{i+1}, \ldots, o_n \rangle \mid (\forall o)(\langle o_1, \ldots, o_{i-1}, o, o_{i+1}, \ldots, o_n \rangle \in ext(\imath^n))\}$.

The condition on $\mathscr{UNIV}_i$ tells us that $\mathscr{UNIV}_i(\imath^n)$ ("the ith-universalization of $\imath^n$") is an $n-1$ place relation which has a given $n-1$ tuple in its extension$_{\mathscr{R}}$ iff for every object o, the n-tuple which results by inserting o in the ith place of the given $n-1$ tuple is in the extension$_{\mathscr{R}}$ of $\imath^n$. Intuitively, $\mathscr{UNIV}_2(\imath^2)$ is the property of bearing $\imath^2$ to everything.

(c) $\mathscr{CONV}_{i,j}$, for each i, j, $1 \leq i < j$, is a function mapping $(\mathscr{R}_j \cup \mathscr{R}_{j+1} \cup \ldots)$ into $(\mathscr{R}_j \cup \mathscr{R}_{j+1} \cup \ldots)$ subject to the condition:
$ext_{\mathscr{R}}(\mathscr{CONV}_{i,j}(\imath^n)) = \{\langle o_1, \ldots, o_{i-1}, o_j, o_{i+1}, \ldots, o_{j-1}, o_i, o_{j+1}, \ldots, o_n \rangle \mid \langle o_1, \ldots, o_i, \ldots, o_j, \ldots, o_n \rangle \in ext_{\mathscr{R}}(\imath^n)\}$.

This says that $\mathscr{CONV}_{i,j}(\imath^n)$ ("the conversion of $\imath^n$ about its ith and j^{th} places") is an n-place relation which has in its extension$_{\mathscr{R}}$ all those n-tuples which result by switching the ith and j^{th} members of every n-tuple in $ext_{\mathscr{R}}(\imath^n)$. So $\langle o_1, o_2 \rangle \in ext_{\mathscr{R}}(\mathscr{CONV}_{1,2}(\imath^2))$ iff $\langle o_2, o_1 \rangle \in ext_{\mathscr{R}}(\imath^2)$.

(d) $\mathscr{REFL}_{i,j}$, for each $i,j, 1 \leq i < j$, is a function mapping $(\mathscr{R}_j \cup \mathscr{R}_{j+1} \cup \ldots)$ into $(\mathscr{R}_{j-1} \cup \mathscr{R}_j \cup \ldots)$ subject to the condition: $ext_{\mathscr{R}}(\mathscr{REFL}_{i,j}(r^n)) = \{\langle o_1, \ldots, o_i, \ldots, o_{j-1}, o_{j+1}, \ldots, o_n\rangle |$ $\langle o_1, \ldots, o_i, \ldots, o_j, \ldots, o_n\rangle \in ext_{\mathscr{R}}(r^n)$ and $o_i = o_j\}$.

When given place numbers i and j, $\mathscr{REFL}_{i,j}(r^n)$ ("the i,j^{th} reflection of r^n") is an $n-1$ place relation which has in its extension$_{\mathscr{R}}$ all those $n-1$ tuples which result by deleting the jth member from every n-tuple in the extension$_{\mathscr{R}}$ of r^n which has identical i^{th} and j^{th} members. This ensures that any object o which falls in the extension$_{\mathscr{R}}$ of $\mathscr{REFL}_{1,2}(r^2)$ is such that o bears r^2 to itself, i.e., $\langle o, o\rangle \in ext_{\mathscr{R}}(r^2)$.

(e) $\mathscr{COND}$ is a function from $(\mathscr{R}_1 \cup \mathscr{R}_2 \cup \ldots) \times (\mathscr{R}_1 \cup \mathscr{R}_2 \cup \ldots)$ into $(\mathscr{R}_2 \cup \mathscr{R}_3 \cup \ldots)$ subject to the condition: $ext_{\mathscr{R}}(\mathscr{COND}(r^n, s^m)) =$ $\{\langle o_1, \ldots, o_n, o'_1, \ldots, o'_m\rangle | \langle o_1, \ldots, o_n\rangle \notin ext_{\mathscr{R}}(r^n)$ or $\langle o'_1, \ldots, o'_m\rangle \in ext_{\mathscr{R}}(s^m)\}$.

$\mathscr{COND}$ maps any n-place relation r^n and m-place relation s^m to an $n+m$ place relation which has in its extension$_{\mathscr{R}}$ any $n+m$ tuple which either fails to have an n-tuple from $ext_{\mathscr{R}}(r^n)$ as its first n members or has an m-tuple from $ext_{\mathscr{R}}(s^m)$ as its second m members. So $\langle o_1, o_2\rangle \in ext_{\mathscr{R}}(\mathscr{COND}(r^1, s^1))$ iff $o_1 \notin ext_{\mathscr{R}}(r^1)$ or $o_2 \in ext_{\mathscr{R}}(s^1)$.

(f) $\mathscr{NEG}$ is a function from $(\mathscr{R}_1 \cup \mathscr{R}_2 \cup \ldots)$ into $(\mathscr{R}_1 \cup \mathscr{R}_2 \cup \ldots)$ subject to the condition: $ext_{\mathscr{R}}(\mathscr{NEG}(r^n)) =$ $\{\langle o_1, \ldots, o_n\rangle | \langle o_1, \ldots, o_n\rangle \notin ext_{\mathscr{R}}(r^n)\}$.

$\mathscr{NEG}(r^n)$ is an n-place relation which has in its extension$_{\mathscr{R}}$ all of the n-tuples not in the extension$_{\mathscr{R}}$ of r^n.

This completes the definitions of the logical functions. They guarantee that the domain of relations, $\mathscr{R}$, houses a rich variety of complex relations.

The fifth member of an interpretation is the last function which imposes a structure on the domains $\mathscr{D}$ and $\mathscr{R}$. We suppose that every property $\mathscr{R}_1$ has an **encoding extension** ("extension$_{\mathscr{A}}$"). The encoding extension of a property is a set of members of $\mathscr{D}$ which encode the property. The fifth member of an interpretation is therefore a function, $ext_{\mathscr{A}}$, which maps $\mathscr{R}_1$ into $\mathfrak{P}(\mathscr{D})$, i.e. $ext_{\mathscr{A}}: \mathscr{R}_1 \to \mathfrak{P}(\mathscr{D})$.

The final member of an interpretation, the $\mathscr{F}$ function, maps the simple names of the language to elements of the appropriate domain. For each object name κ, $\mathscr{F}(\kappa) \in \mathscr{D}$. For each relation name κ^n, $\mathscr{F}(\kappa^n) \in \mathscr{R}_n$. Since "$E!$" is a simple property name, $\mathscr{F}(E!) \in \mathscr{R}_1$, and so $ext_{\mathscr{R}}(\mathscr{F}(E!)) \subseteq \mathscr{D}$. We call this subset of $\mathscr{D}$ the set of **existing** objects ("$\mathscr{E}$"). We call the complement of $\mathscr{E}$ on $\mathscr{D}$ (i.e., $ext_{\mathscr{R}}(\mathscr{NEG}(\mathscr{F}(E!)))$) the set of **abstract** objects ("$\mathscr{A}$").

B. ASSIGNMENTS AND DENOTATIONS

As usual, an assignment with respect to an interpretation $\mathscr{I}$ will be any function, $f_{\mathscr{I}}$, which assigns to each primitive variable an element of the domain over which the variable ranges. And, a denotation function with respect to an interpretation $\mathscr{I}$ and an $\mathscr{I}$-assignment $f_{\mathscr{I}}$, will be any function, $d_{\mathscr{I},f_{\mathscr{I}}}$, defined on the terms of the language, which: (1) agrees with $\mathscr{F}_{\mathscr{I}}$ on the primitive names, (2) agrees with $f_{\mathscr{I}}$ on the primitive variables, and (3) assigns denotations to the complex terms on the basis of the denotations of their parts and the way in which they are arranged. But consider a complex term like "$[\lambda x Px \rightarrow Syx]$". Suppose that $\mathscr{F}_{\mathscr{I}}(P)$ is the property of being a painting and $\mathscr{F}_{\mathscr{I}}(S)$ is the study relation. Our λ-expression would then read: "being an object x such that if x exemplifies paintinghood then y bears the study relation of it". The denotation of this λ-expression will be assigned in terms of the denotations of "P", "S", and "y", and the way in which these parts of the expression are arranged.

Since the denotation function $d_{\mathscr{I},f}$ (for convenience, we drop the subscript on the f) must agree with $\mathscr{F}_{\mathscr{I}}$, we know:

$$d_{\mathscr{I},f}(P) = \mathit{paintinghood}$$
$$d_{\mathscr{I},f}(S) = \text{the } \mathit{study} \text{ relation}$$

$d_{\mathscr{I},f}$ will also agree with f on its assignment to "y"; so let us suppose that $d_{\mathscr{I},f}(y) = o$. However, there are three ways to construct a complex property which might serve to interpret the way in which these simple parts are arranged in the λ-expression. One alternative is to first plug $d_{\mathscr{I},f}(y)$ into the first place of $d_{\mathscr{I},f}(S)$, conditionalize $d_{\mathscr{I},f}(P)$ with the one place property which results, and then reflect the first and second places of the 2-place relation resulting from the conditionalization. This would give us:

$$\mathscr{REFL}_{1,2}(\mathscr{COND}(d_{\mathscr{I},f}(P), \mathscr{PLUG}_1(d_{\mathscr{I},f}(S), d_{\mathscr{I},f}(y)))).$$

On the other hand, we might first conditionalize $d_{\mathscr{I},f}(P)$ with $d_{\mathscr{I},f}(S)$ to get a three place relation, reflect its first and third places to get a 2-place relation, and then plug $d_{\mathscr{I},f}(y)$ into the second place of the result. This would give us:

$$\mathscr{PLUG}_2(\mathscr{REFL}_{1,2}(\mathscr{COND}(d_{\mathscr{I},f}(P), d_{\mathscr{I},f}(S))), d_{\mathscr{I},f}(y)).$$

Finally, we might conditionalize $d_{\mathscr{I},f}(P)$ with $d_{\mathscr{I},f}(S)$, then plug $d_{\mathscr{I},f}(y)$ into the second place of this 3-place relation, and then reflect the first and second places of the result. This would give us:

$$\mathscr{REFL}_{1,2}(\mathscr{PLUG}_2(\mathscr{COND}(d_{\mathscr{I},f}(P), d_{\mathscr{I},f}(S)), d_{\mathscr{I},f}(y))).$$

That is, the following three properties are all sitting around in $\mathscr{R}_1$ and could equally well serve as the denotation of $[\lambda x Px \rightarrow Syx]$ with respect to $\mathscr{I}$ and f:

$$\mathscr{REFL}_{1,2}(\mathscr{COND}(\mathit{paintinghood}, \mathscr{PLUG}_1(\mathit{study}, o)))$$
$$\mathscr{PLUG}_2(\mathscr{REFL}_{1,3}(\mathscr{COND}(\mathit{paintinghood}, \mathit{study})), o)$$
$$\mathscr{REFL}_{1,2}(\mathscr{PLUG}_2(\mathscr{COND}(\mathit{paintinghood}, \mathit{study}), o)).$$

The claim that these three complex properties are in fact the SAME property is a metaphysical thesis of great interest. The idea is that these complicated looking script expressions which are displayed immediately above just represent different decompositions of the same property. Of course, such a thesis needs to be supported, preferably with a (mathematical) theory which predicts when any two such properties or relations are identical.[7] But such a theory has yet to be devised.[8]

Consequently, we face the question, which of the above three properties should be assigned as the denotation$_{\mathscr{I},f}$ of our λ-expression $[\lambda x Px \rightarrow Syx]$? In order to answer this question, we shall develop a mechanical procedure which selects one of the above properties and which makes a similar kind of selection for each of the other λ-expressions. This mechanical procedure is embodied primarily in a classification which partitions the λ-expressions into seven syntactic equivalence classes. Six of these classes will correspond to the logical functions found in $\mathscr{L}$.[9] The seventh houses all of the "simple" λ-expressions. $[\lambda x Px \rightarrow Syx]$ will be categorized as a 1,2-reflection of the expression $[\lambda xu Px \rightarrow Syu]$, which in turn will be categorized as the conditionalization of the two expressions $[\lambda x Px]$ and $[\lambda u Syu]$. The first of these is simple and the second will be categorized as the 1st-plugging of $[\lambda wu Swu]$ by term y. Once the λ-expressions have been partitioned, it will be straightforward to define $\mathscr{I}$-assignment and denotation$_{\mathscr{I},f}$ so that $[\lambda x Px \rightarrow Syx]$ denotes the first of the above three properties. The definitions of $\mathscr{I}$-assignment and denotation$_{\mathscr{I},f}$ follow the partitioning.

Partitioning the λ-expressions. We use μ, ξ, ζ as metavariables ranging over λ-expressions. Suppose μ is an arbitrary λ-expression. Then

$\mu = [\lambda v_1 \ldots v_n \phi]$, for some $\phi, v_1, \ldots, v_n$. Utilizing the following five major rules, we then classify μ as the **i, jth-conversion of** ξ, as the **negation of** ξ, as the **conditionalization of** ξ **and** ζ, as the **ith-universalization of** ξ, as the **i, jth-reflection of** ξ, as the **ith-plugging of** ξ **by** o, or as **elementary.**

(1) If $(\exists i)(1 \leq i \leq n$ and v_i is not the ith free object variable in ϕ and i is the least such number), then where v_j is the ith free object variable in ϕ, μ is the **i, jth-conversion of**

$$[\lambda v_1 \ldots v_{i-1} v_j v_{i+1} \ldots v_{j-1} v_i v_{j+1} \ldots v_n \phi].$$

(2) If μ is not the i, j^{th}-conversion of any λ-expression, then:

(a) if $\phi = (\sim \psi)$, μ is the **negation of** $[\lambda v_1 \ldots v_n \psi]$

(b) if $\phi = (\psi \rightarrow \chi)$, and ψ and χ have no free object variables in common, then where $v_1, \ldots, v_p$ are the variables in ψ and $v_{p+1}, \ldots, v_n$ are the variables in χ, μ is the **conditionalization of** $[\lambda v_1 \ldots v_p \psi]$ **and** $[\lambda v_{p+1} \ldots v_n \chi]$

(c) if $\phi = (\forall v)\psi$, and v is the ith free object variable in ϕ, then μ is the **ith-universalization of** $[\lambda v_1 \ldots v_{i-1} v v_i v_{i+1} \ldots v_n \psi]$.

(3) If μ is none of the above, then if $(\exists i)(1 \leq i \leq n$ and v_i occurs free in more than one place in ϕ and i is the least such number), then where:

(a) k is the number of free object variables between the first and second occurrences of v_i,

(b) ϕ' is the result of replacing the second occurrence of v_i with a new variable v, and

(c) $j = i + k + 1$,

μ is the **i, jth-reflection of** $[\lambda v_1 \ldots v_{i+k} v v_j \ldots v_n \phi']$.

(4) If μ is none of the above, then if o is the left most object term occurring in ϕ, then where:

(a) j is the number of free variables occurring before o,

(b) ϕ' is the result of replacing the first occurrence of o by a new variable v, and

(c) $i = j + 1$,

μ is the **ith-plugging of** $[\lambda v_1 \ldots v_j v v_{j+1} \ldots v_n \phi']$ **by** o.

(5) If μ is none of the above, then

(a) ϕ is atomic,

(b) $v_1, \ldots, v_n$ is the order in which these variables first occur in ϕ,

(c) $\mu = [\lambda v_1 \ldots v_n \rho^n v_1 \ldots v_n]$, for some relation term ρ^n, and

(d) μ is called **elementary**.

Rules (1)–(5) partition the class of λ-expressions into seven equivalence classes. The reader should verify that: $[\lambda x\, Rxb]$ is the 2nd-plugging of $[\lambda xy Rxy]$ by b; $[\lambda x(Px \rightarrow Skx)]$ is the 1, 2-reflection of $[\lambda xy(Px \rightarrow Sky)]$; $[\lambda xy(\forall w)Bxwy]$ is the 2nd-universalization of $[\lambda xwy Bxwy]$; and $[\lambda xy(Rxx \rightarrow Syy)]$ is the conditionalization of $[\lambda x Rxx]$ and $[\lambda y Syy]$; among other examples.

$\mathscr{I}$-assignments.[10] If given an interpretation $\mathscr{I}$ of our language, an **$\mathscr{I}$-assignment**, f, will be any function defined on the primitive variables of the language which satisfies the following two conditions:

(1) where v is any object variable, $f(v) \in \mathscr{D}$
(2) where π^n is any relation variable, $f(\pi^n) \in \mathscr{R}_n$.

Denotations. If given an interpretation $\mathscr{I}$ of our language, and an $\mathscr{I}$-assignment f, we recursively define **the denotation of** term τ **with respect to** interpretation $\mathscr{I}$ **and** $\mathscr{I}$-assignment f ("$d_{\mathscr{I},f}(\tau)$") as follows:

(1) where κ is any primitive name, $d_{\mathscr{I},f}(\kappa) = \mathscr{F}_{\mathscr{I}}(\kappa)$

(2) where v is any object variable, $d_{\mathscr{I},f}(v) = f(v)$

(3) where π^n is any relation variable, $d_{\mathscr{I},f}(\pi^n) = f(\pi^n)$

(4) where $[\lambda v_1 \ldots v_n \rho^n v_1 \ldots v_n]$ is any elementary λ-expression, $d_{\mathscr{I},f}([\lambda v_1 \ldots v_n \rho^n v_1 \ldots v_n]) = d_{\mathscr{I},f}(\rho^n)$

(5) where μ is the ith-plugging of ξ by o, $d_{\mathscr{I},f}(\mu) = \mathscr{PLUG}_i(d_{\mathscr{I},f}(\xi), d_{\mathscr{I},f}(o))$

(6) where μ is the ith-universalization of ξ, $d_{\mathscr{I},f}(\mu) = \mathscr{UNIV}_i(d_{\mathscr{I},f}(\xi))$

(7) where μ is the i,j^{th}-conversion of ξ, $d_{\mathscr{I},f}(\mu) = \mathscr{CONV}_{i,j}(d_{\mathscr{I},f}(\xi))$

(8) where μ is the i,j^{th}-reflection of ξ, $d_{\mathscr{I},f}(\mu) = \mathscr{REFL}_{i,j}(d_{\mathscr{I},f}(\xi))$

(9) where μ is the conditionalization of ξ and ζ, $d_{\mathscr{I},f}(\mu) = \mathscr{COND}(d_{\mathscr{I},f}(\xi), d_{\mathscr{I},f}(\zeta))$

(10) where μ is the negation of $\xi, d_{\mathscr{I},f}(\mu) = \mathscr{NEG}(d_{\mathscr{I},f}(\xi))$.

Here are some examples of λ-expressions and their denotations:

$$d_{\mathscr{I},f}([\lambda x\, Rxa]) = \mathscr{PLUG}_2(d_{\mathscr{I},f}(R), d_{\mathscr{I},f}(a))$$
$$d_{\mathscr{I},f}([\lambda x\, Sxbd]) = \mathscr{PLUG}_2(\mathscr{PLUG}_3(d_{\mathscr{I},f}(S), d_{\mathscr{I},f}(d)), d_{\mathscr{I},f}(b))$$
$$d_{\mathscr{I},f}([\lambda x\, Px \to Skx]) = \mathscr{REFL}_{1,2}(\mathscr{COND}(d_{\mathscr{I},f}(P), \mathscr{PLUG}_1(d_{\mathscr{I},f}(S), d_{\mathscr{I},f}(k))))$$
$$d_{\mathscr{I},f}([\lambda xy (\forall w) Bxwy]) = \mathscr{UNIV}_2(d_{\mathscr{I},f}(B))$$
$$d_{\mathscr{I},f}([\lambda xy\, Rxx \to Syy]) = \mathscr{COND}(\mathscr{REFL}_{1,2}(d_{\mathscr{I},f}(R)), \mathscr{REFL}_{1,2}(d_{\mathscr{I},f}(S)))$$
$$d_{\mathscr{I},f}([\lambda x\, Bx \to (\forall y)(Wyx \to Lmy)]) = \mathscr{REFL}_{1,2}(\mathscr{COND}(d_{\mathscr{I},f}(B), \mathscr{UNIV}_1(\mathscr{REFL}_{1,3}(\mathscr{COND}(d_{\mathscr{I},f}(W), \mathscr{PLUG}_1(d_{\mathscr{I},f}(L), d_{\mathscr{I},f}(m)))))))$$

C. SATISFACTION[11]

If we are given an interpretation $\mathscr{I}$, and an assignment f, we may define f **satisfies** ϕ, recursively, as follows:

(1) If $\phi = \rho^n o_1 \ldots o_n$, f satisfies ϕ iff $\langle d_{\mathscr{I},f}(o_1), \ldots, d_{\mathscr{I},f}(o_n)\rangle \in ext_{\mathscr{R}}(d_{\mathscr{I},f}(\rho^n))$

(2) If $\phi = o\rho^1$, f satisfies ϕ iff $d_{\mathscr{I},f}(o) \in ext_{\mathscr{A}}(d_{\mathscr{I},f}(\rho^1))$

(3) If $\phi = (\sim \psi)$, f satisfies ϕ iff f fails to satisfy ϕ

(4) If $\phi = (\psi \to \chi)$, f satisfies ϕ iff f fails to satisfy ψ or f satisfy χ

(5) If $\phi = (\forall \alpha)\psi$, f satisfies ϕ iff $(\forall f')(f' \overset{\alpha}{=} f \to f'$ satisfies $\phi)$, where: $f' \overset{\alpha}{=} f =_{df} f'$ is an $\mathscr{I}$-assignment just like f except perhaps for what it assigns to α.[12]

D. TRUTH UNDER AN INTERPRETATION

ϕ is **true under** interpretation $\mathscr{I}$ iff every $\mathscr{I}$-assignment f satisfies ϕ. ϕ is **false under** $\mathscr{I}$ iff no $\mathscr{I}$-assignment f satisfies ϕ. Using this definition, we say that ϕ is **valid (logically true)** iff ϕ is true under all interpretations.

The logical axioms which follow in the next section are all valid. We say that an interpretation $\mathscr{I}$ is a **model** of elementary object theory iff all the proper axioms of the theory (Section 4) are true under $\mathscr{I}$.

3. THE LOGIC

The logic for our interpreted language consists of:[13]

A. Logical Axioms.
B. Rules of Inference.

A. THE LOGICAL AXIOMS

There are an infinite number of formulas which are logically true (valid). Some of these are designated as logical axioms and they, together with the rules of inference, store the analytical power of the theory. The logical axioms are introduced by schemata, which indicate that all formulas of a certain form are to be axioms. The schemata fall into three groups: the propositional schemata, the quantificational schemata, and two schemata governing λ-expressions. The second λ-schema will be introduced after some discussion.

Propositional Schemata

LA1: $\phi \rightarrow (\psi \rightarrow \phi)$
LA2: $(\phi \rightarrow (\psi \rightarrow \chi)) \rightarrow ((\phi \rightarrow \psi) \rightarrow (\phi \rightarrow \chi))$
LA3: $(\sim \phi \rightarrow \sim \psi) \rightarrow ((\sim \phi \rightarrow \psi) \rightarrow \phi)$.

Quantificational Schemata

LA4: $(\alpha)\phi \rightarrow \phi^{\tau}_{\alpha}$, where τ is substitutable for α
LA5: $(\alpha)(\phi \rightarrow \psi) \rightarrow (\phi \rightarrow (\alpha)\psi)$, provided α is not free in ϕ.

Lambda Schemata

λ-EQUIVALENCE: Where ϕ is any propositional formula, the following is an axiom:

$$(x_1)\ldots(x_n)([\lambda v_1 \ldots v_n \phi]x_1 \ldots x_n \equiv \phi^{x_1,\ldots,x_n}_{v_1,\ldots,v_n})$$

Here are some instances of λ-EQUIVALENCE:[14]

$$(x)([\lambda y \sim Ry]x \equiv \sim Rx)$$
$$(u)(v)([\lambda xy\, Px \,\&\, Syx]uv \equiv Pu \,\&\, Svu).$$

The first says that an arbitrary object x exemplifies failing-to-exemplify-R iff x fails to exemplify R. The second asserts that any two objects u and v exemplify being-two-objects-such-that-the-first-exemplifies-P-and-the-second-bears-the-S-relation-to-the-first iff u exemplifies P and v bears the S relation to u.

The second λ-schema, λ-IDENTITY, is stated in terms of some defined notation. Its being logically true is a consequence of the fact that given an arbitrary interpretation $\mathscr{I}$ and assignment f, the denotation$_{\mathscr{I},f}$ of an arbitrary relation term ρ^n is identical to the denotation$_{\mathscr{I},f}$ of $[\lambda v_1 \ldots v_n \rho^n v_1 \ldots v_n]$. However, we do not have the primitive logical notion of identity in our object language to express this fact. Nevertheless, we shall designate a logically true formula involving ρ^n and $[\lambda v_1 \ldots v_n \rho^n v_1 \ldots v_n]$ to serve as a definition of identity among relations. This will allow us to interchange ρ^n and $[\lambda v_1 \ldots v_n \rho^n v_1 \ldots v_n]$ in formulas once we add a proper axiom legitimizing substitution of identicals in the next section. The formula which is to serve as the definiens for relation identity itself involves the defined notion of identity among properties. Consider $\boldsymbol{D}_2$:

$$\boldsymbol{D}_2 \qquad F^1 = G^1 =_{df} (x)(xF^1 \equiv xG^1).$$

At first glance, it will not be apparent why this should serve as a good analysis of property identity, but at least it has the merit of not being intuitively false (like a certain other analysis of property identity, namely, that properties are identical iff they are exemplified by the same objects). We will try to provide more justification for this definition in Section 4. Notice that any formula of the form $[\lambda v \rho^1 v] = \rho^1$ will be logically true. In any interpretation, the denotation of the terms flanking the identity sign are identical, so they must have the same encoding extension.

There is a natural way to generalize the definition of property identity to obtain an analysis of relation identity. Consider $\boldsymbol{D}_3$:

$$\begin{aligned}\boldsymbol{D}_3 \qquad & F^n = G^n =_{df} \qquad (\text{where } n > 1)\\ & (x_1)\ldots(x_{n-1})([\lambda y\, F^n y x_1 \ldots x_{n-1}] = [\lambda y\, G^n y x_1 \ldots x_{n-1}]\ \&\\ & \qquad [\lambda y\, F^n x_1 y x_2 \ldots x_{n-1}] =\\ & \qquad [\lambda y\, G^n x_1 y x_2 \ldots x_{n-1}]\ \& \ldots \&\\ & \qquad [\lambda y\, F^n x_1 \ldots x_{n-1} y] = [\lambda y\, G^n x_1 \ldots x_{n-1} y]).\end{aligned}$$

This definition may be read in the following, intuitive manner: relations F^n and G^n are identical iff the one-place properties which result no matter how $n-1$ objects are plugged into them (provided F^n and G^n are plugged up in the same way) are identical (i.e., encoded by the same objects). Given this definition, formulas of the form $[\lambda v_1 \ldots v_n \rho^n v_1 \ldots v_n] = \rho^n$ are logically true (and I do not see a way to derive them). So we officially add λ-IDENTITY as a logical axiom schema governing our structure:

λ-IDENTITY: where ρ^n is any relation term and $v_1, \ldots, v_n$ are any object variables, the following is an axiom:

$$[\lambda v_1 \ldots v_n \rho^n v_1 \ldots v_n] = \rho^n.$$

B. RULES OF INFERENCE

(1) Arrow Elimination ("$\rightarrow$ E"): from ϕ and $\phi \rightarrow \psi$, we may infer ψ

(2) Universal Introduction ("UI"): from ϕ, we may infer $(\alpha)\phi$.

Officially, these are all the rules we'll need. Open formulas are assertible. We define **the universal closure of** $\phi(\alpha_1, \ldots, \alpha_n)$ to be $(\alpha_1)\ldots(\alpha_n)\phi$. It is straightforward to show that open formulas are true iff their universal closures are. In the usual manner, a **proof** will be any finite sequence of formulas $\phi_1, \ldots, \phi_n$ such that, for each i, either ϕ_i is a logical axiom or ϕ_i may be inferred from some of the preceding formulas by a rule of inference. ϕ is a **theorem of logic (logical theorem)** iff there is a proof of which ϕ is the last member. We write "$\vdash \phi$" to indicate that ϕ is a theorem of logic. ϕ is **a proof-theoretic consequence of (derivable from, provable from)** a set Γ of formulas iff there is a sequence of formulas $\phi_1, \ldots, \phi_n$ such that $\phi = \phi_n$ and, for each i, either (a) ϕ_i is in Γ, or (b) ϕ_i is a logical axiom, or (c) ϕ_i may be inferred from some of the preceding formulas by a rule of inference. Such a sequence is called a proof of ϕ from Γ and to indicate this, we write either "$\vdash_\Gamma \phi$" or "$\Gamma \vdash \phi$".

If the set of formulas Γ constitute the proper axioms of some theory, and if $\Gamma \vdash \phi$ but it is not the case that $\vdash \phi$. then we say that ϕ is a **proper theorem** of Γ. It is important to distinguish the logical theorems of a theory from the proper theorems. The logical theorems are derivable from the logical axioms and rules of inference alone, whereas the proper theorems depend on one of the proper axioms. We may define this notion of dependence as follows. Let ψ be a formula in a set Γ of formulas. Assume that we are given a proof $\phi_1, \ldots, \phi_n$ from Γ, together with the justification for each step of the proof. We then say ϕ_i **depends upon** ψ in this proof iff: (a) ϕ_i is ψ, and the justification for ϕ_i is that it belongs to Γ, or (b) ϕ_i is justified as a direct consequence by $\rightarrow$E or UI of some preceding formulas of the sequence, where at least one of these preceding formulas depends on ψ.

It will be convenient to employ the many standard derived rules of inference. For example, we call the rule of inference derivable from → E and LA4 **universal elimination** ("UE"). Standard formulations of the existential introduction and elimination rules ("EI" and "EE"), the quantifier negation rules ("QN"), and the introduction and elimination rules for $\sim, \&, v,$ and $\equiv$ are employed. And we shall use conditional and indirect proof techniques. The proofs sketched in the text are all constructed with the aid of these derived rules and proof techniques.

By using UE on the universal quantifiers of the instances of λ-EQUIVALENCE, we obtain biconditionals. Rules of inference governing the biconditional allow us to introduce (eliminate) λ-expressions into proofs when the right (left) side of the biconditional is added as a premise. We may shorten this procedure by formulating two rules of inference derived from λ-EQUIVALENCE, λI and λE: where ϕ is any propositional formula with object terms $o_1, \ldots, o_n$, and $v_1, \ldots, v_n$ are object variables substitutable for $o_1, \ldots, o_n$, respectively, then the following are rules of inference:

(1) λ-Introduction ("λI"): from ϕ, we may infer

$$[\lambda v_1 \ldots v_n \phi^{v_1, \ldots, v_n}_{o_1, \ldots, o_n}] o_1 \ldots o_n.$$

(2) λ-Elimination ("λE"): from $[\lambda v_1 \ldots v_n \phi^{v_1, \ldots, v_n}_{o_1, \ldots, o_n}] o_1 \ldots o_n$, we may infer ϕ.

Also, since $[\lambda v_1 \ldots v_n \phi]$ is an n-place relation term, it is subject to existential introduction. We get an important logical theorem schema by applying EI to λ-EQUIVALENCE:

THEOREM(S) ("RELATIONS"): where ϕ is a propositional formula which has no free F^n's, but has $x_1, \ldots, x_n$ free, the following is a theorem:

$$(\exists F^n)(x_1)\ldots(x_n)(F^n x_1 \ldots x_n \equiv \phi).$$

The instances of this schema tell us what complex properties and relations there are. Here are some examples:

(a) $(y)(\exists F)(x)(Fx \equiv Gyx)$ (by UI)

(b) $(\exists F)(x)(Fx \equiv \sim Gx)$

(c) $(\exists F)(x)(Fx \equiv Gx \;\&\; Hx)$

(d) $(\exists F)(x)(Fx \equiv Gx \vee Hx)$

(e) $(\exists F)(x)(y)(Fxy \equiv Gyx)$

(f) $(\exists F)(x)(Fx \equiv (\forall y)Gxy)$.

Axioms (a)–(f) assert, respectively, that for every object y and two place relation G, there is a property which results by plugging y into the first place of G, that every property has a negation, every two properties have a (non-disjoint) conjunction and disjunction, every two place relation has a converse, every two place relation has a universalization on its second place.

Note that RELATIONS, $\boldsymbol{D}_2$, and $\boldsymbol{D}_3$ jointly constitute a full-fledged theory of relations. We no longer need to suppose that relations are "creatures of darkness". They have precise "being" conditions and precise identity conditions. It is not a consequence of our theory that equivalent relations are identical; we cannot prove $(F^n)(G^n)(x_1)\ldots(x_n)((F^n x_1 \ldots x_n \equiv G^n x_1 \ldots x_n) \rightarrow F^n = G^n)$. So it does not follow from the fact that being a rational animal and being a featherless biped are exemplified by exactly the same objects that these two properties are identical.

We should also note that there are two senses of "F is a equivalent to G" when F and G are properties. One sense is that F and G are exemplified by the same objects. The second sense is that F and G are encoded by the same objects. We have stipulated that properties equivalent in the latter sense are identical. In what follows, we always use "equivalent" in the former sense.

We call the slightly modified second order language, together with its semantics and logic, the **object calculus (with complex relation terms)**. The object calculus is the formal system in which the proper axioms of the elementary theory of abstract objects are stated.

4. THE PROPER AXIOMS[15]

We have now embedded our primitive metaphysical notions in the atomic formulas of the language and embedded our primitive logical notions in the complex formulas and terms of the language. To state the theory of abstract objects, we shall also need to use our two primitive theoretical relations, ***existence*** and ***E-identity***. The theory has four axioms, two of which are schemata and which involve defined notions. Though these axioms are not logically true, we nevertheless suppose them to be true A PRIORI. The first two axioms are non-schematic and express truths about existing objects. The first schema utilizes all the defined notions of identity constructed in both Sections 3 and 4 and tells us about the behavior of any entities which satisfy the definitions. The second schema tells us what abstract objects there are. Since the schemata indicate that all sentences of a certain form are to be axioms, we end up with a denumerably infinite number of proper axioms. These axioms, plus the definitions in terms of which they are stated, constitute the first principles of the elementary theory of abstract objects.

The first axiom tells us that two objects bear the identity$_E$ relation to one another iff they both exist and exemplify the same properties:[16]

AXIOM 1. ("*E*-IDENTITY"): $x =_E y \equiv E!x \,\&\, E!y \,\&\, (F)(Fx \equiv Fy)$.

The second axiom tells us that no existing objects encode properties:

AXIOM 2. ("NO-CODER"): $E!x \rightarrow \sim(\exists F)xF$.

The theory does not assert that there are any existing objects. Instead, these first two axioms are meant to capture natural assumptions we make about existing objects, should there be any.

In a sense, our first axiom tells us the conditions under which existing objects are identical. Recall that $\boldsymbol{D}_1$ (Section 1) says that abstract objects are objects which exemplify the property of non-existence. Since this partitions the domain of objects into disjoint classes, the following definition is a completely general definition of object identity:

$$\boldsymbol{D}_4 \qquad x = y \;=_{df} x =_E y \vee (A!x \,\&\, A!y \,\&\, (F)(xF \equiv yF)).$$

E-IDENTITY, $\boldsymbol{D}_2$, $\boldsymbol{D}_3$, and $\boldsymbol{D}_4$ allow us to prove one of the laws of identity as a theorem schema:

THEOREMS ("IDENTITY INTRODUCTION"): $\alpha = \alpha$, where α is any variable.

Proof. If α is an object variable x and $E!x$, then since we have $(F)(Fx \equiv Fx)$ from propositional logic and UI, we may use *E*-IDENTITY to prove $x =_E x$. So $x = x$, by $\boldsymbol{D}_4$. If $\sim E!x$, then x is abstract and similar techniques get us the right hand disjunction of $\boldsymbol{D}_4$. If α is a one-place property variable F^1, we easily get $(x)(xF^1 \equiv xF^1)$. So by $\boldsymbol{D}_2$, $F^1 = F^1$. And a generalized version of this procedure gets us $F^n = F^n$. ☒

In what follows, we abbreviate "IDENTITY INTRODUCTION" as "$=$I."

We may complete the presentation of our THEORY of identity by introducing the third axiom of the theory of abstract objects. Since all of the definienda in $\boldsymbol{D}_2$, $\boldsymbol{D}_3$, and $\boldsymbol{D}_4$ have the form $\alpha = \beta$, we assert that the following axiom is true:

AXIOM 3. ("IDENTITY"): $\alpha = \beta \rightarrow (\phi(\alpha, \alpha) \equiv \phi(\alpha, \beta))$, where $\phi(\alpha, \beta)$ is the result of replacing some, but not necessarily all, free occurrences of α by β in $\phi(\alpha, \alpha)$, provided β is substitutable for α in the occurrences of α it replaces.[17]

The rule of inference derivable from $\rightarrow$E and IDENTITY is called **identity elimination** ("$=$E").

The schema for abstract objects generates the most important set of axioms of the theory. In effect, the schema guarantees that for every expressible set of properties, there is an abstract object which encodes just the members of the set.[18] However, the schema does this without a commitment to sets. We generally use open formulas with one free property variable with this axiom, though sentences will work as well (they express vacuous conditions). Metalinguistically, it is legitimate to talk about the set of properties satisfying a given condition, but in the object language, our schema says something more like: for every condition on properties, there is an abstract object which encodes just the properties which meet the condition:[19]

AXIOM(S) 4. ("*A*-OBJECTS"): for any formula ϕ where x is not free, the following is an axiom:

$$(\exists x)(A!x \;\&\; (F)(xF \equiv \phi)).$$

Some examples will help. If we let "$F = R \vee F = S$" be our formula ϕ, and suppose that "R" denotes roundness and "S" denotes squareness, then our axiom guarantees that there is a "round square" as follows:

$$(\exists x)(A!x \;\&\; (F)(xF \equiv F = R \vee F = S)).$$

Suppose a_0 is such an object. It is easy to see that a_0 must be unique. For suppose some other distinct abstract object, say a_1, encoded exactly roundness and squareness. By $\boldsymbol{D}_4$, it would follow that either a_1 encoded a property a_0 did not, or vice versa, contrary to hypothesis.

Let us use the standard notation "$(\exists!x)\psi$" ("there is a unique x such that ψ") to abbreviate $(\exists x)(\psi \;\&\; (y)(\psi^y_x \rightarrow y = x))$. Then, given $\boldsymbol{D}_4$, the following is a consequence of *A*-OBJECTS:

THEOREM(S) ("UNIQUENESS"): for any formula ϕ where x is not free, the following is a theorem:

$$(\exists!x)(A!x \;\&\; (F)(xF \equiv \phi)).$$

Proof. An arbitrary instance of A-OBJECTS asserts that there is an abstract object which encodes EXACTLY the properties which satisfy the given formula. But there could not be distinct such objects, since distinct abstract objects must differ with respect to at least one of the properties they encode. ☒

Another instance of the schema for objects says that there is an "existent golden mountain". Suppose "G" denotes goldenness and "M" denotes mountainhood. We then have:

$$(\exists x)(A!x \;\&\; (F)(xF \equiv F=G \vee F=M \vee F=E!)).$$

It follows that there is an abstract object which encodes a property it fails to exemplify. It is a contingent fact that there does not exist an object which exemplifies all the properties that this object encodes.

By letting $\phi = \ulcorner F \neq F \urcorner$, we obtain the empty object – it fails to encode any properties. By letting $\phi = \ulcorner F = F \urcorner$, we obtain the universal object – it encodes every property.

Suppose "a_5" denotes Socrates. Then the following instance of A-OBJECTS yields an A-object which encodes exactly the properties Socrates exemplifies:

$$(\exists x)(A!x \;\&\; (F)(xF \equiv Fa_5)).$$

We might call this object Socrates' ***blueprint***, and call Socrates the ***correlate*** of the blueprint.[20] We define these terms as follows:[21]

D$_5$ $\quad$ *x is the blueprint of y* and *y is the correlate of x* ("*Blue* (x, y)" and "*Cor*(y, x)") $=_{df} (F)(xF \equiv Fy)$.

A-OBJECTS guarantees that every object, existing or abstract, has a unique blueprint:

$$(y)(\exists!x)(A!x \;\&\; (F)(xF \equiv Fy)).$$

This follows by UI on the instance of UNIQUENESS which results when "Fy" is the formula ϕ.

Given any object b, A-OBJECTS yields an object which encodes all the properties b fails to exemplify. Given any two objects b and c, A-OBJECTS yields an object which encodes (1) just the properties b and c have in common, (2) just the properties exemplified by either b or c, and (3) just the relational properties b has with respect to c. This last object is yielded by the following instance:

$$(\exists x)(A!x \,\&\, (F)(xF \equiv (\exists G^2)((Gbc \,\&\, F=[\lambda x\, Gxc]) \vee (Gcb \,\&\, F=[\lambda x\, Gcx])))).$$

These examples give one a pretty good idea of what A-OBJECTS says.

We use A-OBJECTS to justify the definition we have proposed for property identity. Suppose that instead of defining identity between properties as in $\boldsymbol{D}_2$, we had added primitive identity formulas between property terms. Then the following would have been a consequence of A-OBJECTS:

$$(G)(H)((x)(xG \equiv xH) \rightarrow G=H).$$

To see this, suppose arbitrary properties P and Q were encoded by exactly the same objects, but were distinct. By A-OBJECTS, it would have followed that there is an object which encodes just P, without encoding Q, contrary to hypothesis.

Since we would have had this consequence had property identity been primitive, there was every reason to just define identity among properties. Semantically, our definition ensures that two properties which have the same extension$_{\mathscr{A}}$ are identical. But as we have seen our theory does not commit us to the view that properties exemplified by the same objects (i.e., which have the same extension$_{\mathscr{R}}$) are identical. An overriding reason for choosing the style of semantics we have employed is that properties and relations are not identified with their extensions$_{\mathscr{R}}$. The semantics does not force upon us a view to which the theory is not committed.

E-IDENTITY, NO-CODER, IDENTITY, and A-OBJECTS are jointly called **the elementary theory of abstract objects.** Good evidence for thinking that the theory is consistent may be found in Appendix A, parts A and B, where the reader will find a full discussion of the solutions to the paradoxes which have been avoided as well as a model in Zermelo–Fraenkel set theory of the monadic portion of our theory. One cannot just model A-objects as sets of ordinary properties. That is because sets of ordinary properties cannot exemplify the very same properties which serve as their elements. That would be a violation of type, just as in ZF, no set of sets can be an element of one of its members. But in object theory, A-objects may exemplify the very same properties which they encode. For example, any A-object which encodes $[\lambda x \sim E!x]$ also exemplifies this property.[22] And in the next section, we will try to harness an entire range of intuitions about which other properties A-objects exemplify. These will be properties drawn from the same stock of properties which A-objects encode.

Before we turn to this next section, there is an interesting consequence of the theory of which the reader should be warned. Some complex relations do not have unique constituents. Specifically, it is provable, given any two place relation R, that for some objects a_3 and $a_4, a_3 \neq a_4$, that $[\lambda y Rya_3] = [\lambda y Rya_4]$. Here is how:

Let R be any two place relation. By A-OBJECTS, $(\exists x)(A!x \,\&\, (F)(xF \equiv (\exists u)(F = [\lambda y Ryu] \,\&\, \sim uF)))$. Call this object a_3 and suppose that a_3 fails to encode $[\lambda y Rya_3]$. So by definition of a_3, $\sim (\exists u)([\lambda y Rya_3] = [\lambda y Ryu] \,\&\, \sim u[\lambda y Rya_3])$. That is, $(u)([\lambda y Rya_3] = [\lambda y Ryu] \rightarrow u[\lambda y Rya_3])$. So by instantiating to a_3, it follows that a_3 does encode $[\lambda y Rya_3]$, contrary to hypothesis. So suppose a_3 encodes $[\lambda y Rya_3]$. Then $(\exists u)(\lambda y Rya_3] = [\lambda y Ryu] \,\&\, \sim u[\lambda y Rya_3])$. Call such an object a_4. So $[\lambda y Rya_3] = [\lambda y Rya_4]$, but $a_3 \neq a_4$, since a_3 encodes $[\lambda y Rya_3]$ and a_4 does not. ⊠[23]

The reader is asked to postpone judgement about the seriousness of this result until after the applications have been considered.

5. AN AUXILIARY HYPOTHESIS

In Chapter II, we shall put the theory we've now formulated to work. For these applications, we add to our primitive vocabulary abbreviations of the "gerundive versions" of standard English transitive verbs, intransitive verbs, predicate adjectives, and predicate nouns.[24] By the "gerundive version" of these words, I mean the phrases constructed out of English gerunds which can appear in the subject places or direct object places of English sentences. Here are some examples:

A. Transitive verbs	*Gerundive Version*
kick	(the) kicking (relation)
worship	(the) worshipping (relation)
hate	(the) hating (relation)
B. Intransitive verbs	
run	(the property of) running
walk	(the property of) walking

C. Predicate adjectives

red	(the property of) being red
courageous	(the property of) being courageous
happy	(the property of) being happy

D. Predicate nouns

horse	(the property of) being a horse
person	(the property of) being a person
building	(the property of) being a building

We abbreviate these gerundive versions in our language with single letters, appropriately chosen. For example, "K" might abbreviate "the kicking relation", etc. (Also, we shorten our readings of λ-expressions so that they more closely resemble their natural language counterparts. So instead of reading "$b[\lambda x\, Cx \;\&\; Px]$" as "$b$ encodes being an object x such that x exemplifies being courageous and x exemplifies being a person", we read it as "b encodes being a courageous person".)

These additions to our primitive vocabulary are supposed to reveal our pretheoretic conceptions about what simple properties and relations there are. By adding these properties and relations to our system, *A*-OBJECTS provides us with a rich variety of abstract objects which encode familiar sorts of simple and complex properties.

These additions also make it possible to state an auxiliary hypothesis of the elementary theory – an hypothesis to which we shall appeal on occasion in the applications. Despite its rather vague character, it grounds a wide range of intuitions some of us may share about abstract objects. Pretheoretically, we have a pretty good idea of what properties existing objects exemplify. And the theory tells us the conditions under which both existing and abstract objects encode these properties. But other than being abstract (i.e., $[\lambda x \sim E!x]$), we have not said anything about which properties abstract objects exemplify.

Some of us may share the following intuitions. Abstract objects DO NOT EXEMPLIFY the following properties: being round, having a shape, being red, having a color, being large, having a size, being soft, having a texture, having mass, having spatio-temporal location, being visible, being capable of thought (this is NOT to say that they are not thought of), being capable of feeling, etc. In addition, it might seem that no two abstract objects could ever meet each other, kick each other, kiss each other, etc. I am sure the reader can provide many more examples. If these intuitions are true, then by λ-EQUIVALENCE, A-objects exemplify the negations of these properties and relations. I think we have intuitions to this effect as well.

These properties and relations are "ordinary" properties and relations of existing objects. Mally, Meinong, Findlay, Parsons, and others call them "nuclear" relations (Routley calls them "characterizing" relations). They are to be distinguished from "extranuclear" relations such as being abstract, being thought about, being written about, being worshipped, being more famous than, etc. Note that many of these are "intentional" relations. We can easily imagine that abstract objects exemplify these extranuclear relations.[25]

We shall not pursue this distinction among relations in any detail. We mention it because there will be occasion to appeal to the above intuitions and it would be nice to ground them all in some general principle. Consequently, we suppose that no abstract objects exemplify nuclear relations.

We incorporate this hypothesis into elementary object theory by assuming that we can somehow characterize nuclearity in our language.[26] Suppose that for each $n, n \geq 1$, "Nuclear (F^n)" is some open formula with one free n-place relation variable F^n. We then stipulate:

AUXILIARY HYPOTHESIS: $(F^n)(x_1)\dots(x_n)(A!x_1 \;\&\; \dots \;\&\; A!x_n \;\&\;$ Nuclear $(F^n) \rightarrow \sim F^n x_1 \dots x_n)$.

We trouble the reader with this hypothesis because it seem likely that some such set of truths like these govern abstract objects A PRIORI.

APPLICATIONS OF THE ELEMENTARY THEORY

It will be important to have definite descriptions (which involve the primitive logical notion ***the***) to facilitate the following applications. Let us stipulate that where ϕ is any formula with one free x-variable, $(\iota x)\phi$ ("the object x such that ϕ") is to be a complex object term of our language. Some examples might be: $(\iota y)(E!y \mathbin{\&} Typ)$ ("the object which exists and taught Plato") and $(\iota x)(A!x \mathbin{\&} (F)(xF \equiv F = R \vee F = S))$ ("the object which encodes just roundness and squareness"). Semantically, we interpret descriptions like $(\iota x)\phi$ as denoting the unique object which satisfies ϕ, if there is one, and as not denoting anything if there is not one. To guarantee that descriptions work in our system just as we would expect them to A PRIORI, we add a proper axiom schema which asserts that atomic formulas or defined identity formulas ψ in which there occurs a description $(\iota x)\phi$ are true iff there is a unique object satisfying ϕ and there is something which satisfies both ϕ and ψ.[1]

DESCRIPTIONS: where ψ is any atomic formula or defined object identity formula with one free object variable v, the following is a proper axiom:

$$\psi_v^{(\iota x)\phi} \equiv (\exists! y)\phi_x^y \mathbin{\&} (\exists y)(\phi_x^y \mathbin{\&} \psi_v^y).$$

Here are a few examples:

$$E!(\iota x)Txp \equiv (\exists! y)Txy \mathbin{\&} (\exists y)(Txy \mathbin{\&} E!y)$$

$$b = (\iota x)Px \equiv (\exists! y)Py \mathbin{\&} (\exists y)(Py \mathbin{\&} b = y).$$

The first of these might say: the teacher of Plato exists iff there is a unique teacher of Plato and something which is a teacher of Plato exists. The second says: object b is identical to the object exemplifying P iff there's a unique thing exemplifying P and something exemplifying P is identical with b. Given DESCRIPTIONS, we may easily derive instances of the following schema: $\tau = (\iota x)\phi \rightarrow \phi_x^\tau$ (let $\psi = \ulcorner \tau = v \urcorner$; then $\psi_v^{(\iota x)\phi} = \ulcorner \tau = (\iota x)\phi \urcorner$, which is the left side of DESCRIPTIONS).

There are modifications and restrictions that must be incorporated into the definitions of Chapter I in order to accommodate terms that might fail to denote. A detailed description of these appears in the Appendix to this chapter and interested readers are directed there. For the most part, these modifications and restrictions will not affect what follows, since we shall utilize only descriptions which provably have denotations. So the above discussion is all that is necessary for understanding the following applications.

1. MODELLING PLATO'S FORMS[2]

In this section, we construe certain assertions by Plato as consequences of the theory. Most philosophers today regard Plato's Forms as first level properties of some sort and view participation as just exemplification. But this view of Plato from within Russellian background theory turns Plato's major principle about the Forms into a triviality.

Plato's major principle about the Forms is the One Over the Many Principle. It is stated principally in Parmenides (132a).[3] The following characterization is, I think, a faithful one:

(OMP) If there are two distinct F-things, then there is a Form of F in which they both participate.

According to the orthodox view,

The Form of $F =_{df} F$
x participates in $F =_{df} Fx$.

So translating (OMP) into a standard second order predicate calculus, we would get:

$$x \neq y \,\&\, Fx \,\&\, Fy \rightarrow (\exists G)(G = F \,\&\, Gx \,\&\, Gy).$$

But the consequent of this conditional just follows from the antecedent by existential introduction. Clearly, we do not want to attribute such a triviality to Plato.[4] Yet it is difficult to conceive of it as an interesting metaphysical truth from within the Russellian framework.

In object theory, however, we may think of Forms as just a special kind of A-object. When (OMP) is translated into our language, it turns out to be an interesting theorem. To see this, consider the following series of definitions and proofs:[5]

$\boldsymbol{D}_6$ *x is a Form of G* ("*Form* (x, G)") $=_{df} A!x \,\&\, (F)(xF \equiv F = G)$.

So a Form of G is any abstract object which encodes just G. So we have:

THEOREM 1. $(G)(\exists x)Form(x,G)$.
Proof. By A-OBJECTS.

In fact, given UNIQUENESS (I., Section 4), it also follows that:

THEOREM 2. $(G)(\exists !x)Form(x,G)$.

Given Theorem 2, we know that the description $(\iota x)(A!x \,\&\, (F)(xF \equiv F=G))$ ("the Form of G") always has a denotation. For convenience, let us use "Φ_G" to abbreviate $(\iota x)(A!x \,\&\, (F)(xF \equiv F=G))$. We then have as a simple consequence of DESCRIPTIONS:

THEOREM 3. $\Phi_G G$ ("The Form of G encodes G").
Proof. By DESCRIPTIONS, $\Phi_G G \equiv (\exists !y)(A!y \& (F)(yF \equiv F=G))$ & $(\exists y)(A!y \,\&\, (F)(yF \equiv F=G) \,\&\, yG)$. The right side of this biconditional is easily obtainable from Theorem 2. ☒

Now we can define participation:

D_7 *y participates in x* ("$Part(y,x)$") $=_{df} (\exists F)(xF \,\&\, Fy)$.

So something participates in the Form of G just in case there's a property the Form encodes which the object exemplifies. All objects which exemplify redness participate in the Form of Redness ("Φ_R").

These definitions validate (OMP). The translation of (OMP) into our language turns out to be a theorem:

THEOREM 4. $x \neq y \,\&\, Fx \,\&\, Fy \rightarrow (\exists u)(u=\Phi_F \,\&\, Part(x,u) \,\&\, Part(y,u))$.
Proof. Assume $a \neq b$, Pa, and Pb, where a,b are arbitrary objects and P is an arbitrary property. By $=$I, we have $\Phi_P = \Phi_P$.[6] By Theorem 3 and the above assumptions, we have $\Phi_P P \,\&\, Pa$. So $(\exists G)(\Phi_P G \& Ga)$, i.e., $Part(a,\Phi_P)$. By the same reasoning, $Part(b,\Phi_P)$. So $\Phi_P = \Phi_P \,\&\, Part(a,\Phi_P) \,\&\, Part(b,\Phi_P)$. So, $(\exists u)(u=\Phi_P \& Part(a,u)$ and $Part(b,u))$. ☒

Another theorem quickly falls out of these definitions:

THEOREM 5. $Fx \equiv Part(x,\Phi_F)$.
Proof. ($\rightarrow$) Assume Fx. By Theorem 3, $Part(x,\Phi_F)$. ($\leftarrow$) Assume

$Part(x, \Phi_F)$. Call the property Φ_F encodes and x exemplifies, G. Since Φ_F encodes just F, it must be that $G=F$. So Fx. ☒

So in our system, the notions of exemplification and participation are distinct (unlike the orthodox view) though nonetheless equivalent. This should preserve at least some of the intuitions of orthodox theorists.

On our theory, some Forms participate in other Forms, and indeed, some Forms participate in themselves. Consider the property $[\lambda x \sim E!x]$ ("$\bar{E}!$"). Let us call this property: ***Platonic existence***. Since all A-objects fail to exist, they all exemplify Platonic existence. In particular, we have:

THEOREM 6. $(x)((\exists F)(x=\Phi_F) \rightarrow \bar{E}!x)$.

So the Forms exemplify a kind of existence which is different from the existence exemplified by actual objects.[7] But now consider $\Phi_{\bar{E}!}$, which we may call ***Platonic Being***, or ***Reality***. From Theorems 5 and 6 it follows that:

THEOREM 7. $(x)((\exists F)(x=\Phi_F) \rightarrow Part(x, \Phi_{\bar{E}!}))$.

So all Forms participate in Platonic Being.[8] In particular, $\Phi_{\bar{E}!}$ participates in itself, justifying our claim that some Forms participate in themselves.

To reach this conclusion, we might also have used the AUXILIARY HYPOTHESIS and the assumption that being blue, for example, is a nuclear property. It would follow that all A-objects fail to exemplify this property. So all A-objects would exemplify $[\lambda x \sim Bx]$("$\bar{B}$"), where "B" denotes being blue. Then by Theorem 5, all Forms participate in $\Phi_{\bar{B}}$. So would $\Phi_{\bar{B}}$.

Consider now the Third Man Argument. This is a puzzle which commentators say Plato produces in the *Parmenides* (132aff.).[9] The puzzle is that several of Plato's principles about the Forms seem to be jointly inconsistent. We have seen two of these principles: (OMP) (Theorem 4) and the Uniqueness Principle (Theorem 2). There are two others: the Self-Predication Principle and the Non-Identity Principle:

(SP) The Form of F is F

(NI) If something participates in the Form of F, then it is not identical with that Form.

We can prove a contradiction if we assume that there are two distinct F-things x and y. By (OMP), there is a Form of F in which x and y participate. By (SP), the Form of F is an F-thing. By (NI), it is distinct from x and y. But then, (OMP) guarantees that there is another Form of F in which x and the first Form participate. Then (NI) yields the conclusion that the latter Form must be distinct from the first. But this violates the uniqueness principle, which says that the Form of F is unique.

On the theory we have presented, (NI) must be false. We can derive its negation as a theorem:

THEOREM 8. $\sim(x)(Part(x, \Phi_F) \rightarrow x \neq \Phi_F)$.
Proof. Consider $\Phi_{\bar{E}!}$. ☒

So by rejecting (NI), we dissolve the puzzle.

However, it is worthwhile to examine (SP). If we translate (SP) into our language as Φ_F exemplifies F, then it must be false. This time, consider $\Phi_{E!}$. But if we translate (SP) into our language as Φ_F encodes F, then it turns out to be Theorem 3. Does the word "is" in the (SP) principle mesh the distinction between exemplifying and encoding a property?

Of course we cannot generalize on this one example, but we can look for further evidence for thinking that the "is" of English is ambiguous. Maybe we have an option of translating a sentence involving the predicative "is" as either an exemplification or an encoding formula. And in case there is such an ambiguity, let us now stipulate that whenever we use the word "is" in its predicative sense in what follows, we shall mean "exemplifies".[10]

We may conclude, with respect to the Third Man Argument, that our theory rules that (OMP) and (U) (Uniqueness Principle) are true, that (NI) is false, and that (SP) has a true reading and a false one. Since we abandon the (NI) principle, further research should be directed toward the question of how deeply Plato was committed to it.

Finally, we discuss the *Sophist*. The four assertions by Plato in that work that we discuss are ones which, taken together, are somewhat mysterious. Many scholars regard Plato's theory of Forms as his attempt to reconcile two major philosophical schools of thought. The first was the school of Parmenides, founded as the view that the world had to be considered as a whole without parts, without motion and change, and without generation and decay. The opposing school (Thales, Anaximander, Anaximenes, Heraclitus, Empedocles, and the Atomists) denied this and attempted to isolate the elementary parts of the world, the interaction of which was responsible for motion, change, generation, and decay. Plato's Forms were entities he postulated to capture certain truths of the Parmenidean school – they were changeless, motionless, and eternal. Yet Plato allowed that there were ordinary objects which moved, changed, came into being, and passed away. But, apparently, he supposed them to have a lesser degree of reality.

Plato's attempt to capture the Parmenidean truths was not completely successful. Some Forms gave him trouble, especially the ones which reflected some of the more mundane things in the world. He could never quite accept the fact that there were Forms with respect to hair, dirt, or mud. And the Form of motion – did it move? If so, how could it remain a Form? Forms were supposed to be motionless. Given the (SP) principle, how could there be a real Form of Motion if it did not move? And how do the Forms of Motion and Rest interact with each other?

In this context, the following four assertions by Plato in the *Sophist* seem mysterious:

(1) Rest and Motion are completely opposed to one another (250a).

(2) Rest and Motion are real (250a).

(3) Reality must be some third thing (250b).

(4) In virute of its own nature, then, reality is neither at rest nor in movement (250c).

To analyze these assertions, we need the following definitions and (reasonable) assumptions, where "*M*" denotes being in motion.

$\boldsymbol{D}_8$ *Being at rest* ("*R*") $=_{df} [\lambda x \sim Mx]$

$\boldsymbol{A}_1$ $Nuclear(M)$

$\boldsymbol{A}_2$ $M \neq R \;\&\; M \neq \bar{E}! \;\&\; R \neq \bar{E}!$.

(1) may be interpreted as a true statement about the Forms. Consider (1a):

(1a) $(x)(Part(x, \Phi_M) \equiv \sim Part(x, \Phi_R))$.

This is provable, given $\boldsymbol{D}_8$, λ-EQUIVALENCE, and Theorem 5. That is, by $\boldsymbol{D}_8$ and λ-EQUIV, something exemplifies being in motion iff it fails to exemplify being at rest. So by Theorem 5, something participates in Φ_M iff it fails to participate in Φ_R.

There is also an uncharitable way to interpret (1) as a statement about the Forms. Consider (1b):

(1b) $\sim R\Phi_M \ \& \sim M\Phi_R$ ("The Form of Motion does not exemplify being at rest and the Form of Rest does not exemplify being in motion").

This is false, since by A_1, D_8, the AUXILIARY HYPOTHESIS, and λ-EQUIV, the Form of Motion does exemplify the negation of a nuclear property.

Consider (2), "Rest and motion are real". (2a) seems to be a good candidate for translating it:

(2a) $\bar{E}!\Phi_R \ \& \ \bar{E}!\Phi_M$ ("The Forms of Rest and Motion exemplify Platonic existence").

(2a) is a theorem. We also know that both Φ_R and Φ_M participate in $\Phi_{\bar{E}!}$. If we define "blend with" as "participate in", we get that both of these Forms blend with Being or Reality. (3) could be read as (3a):

(3a) $\Phi_R \neq \Phi_{\bar{E}!} \ \& \ \Phi_M \neq \Phi_{\bar{E}!}$.

This is provable from assumption A_2.

Finally, we consider (4), "In virtue of its own nature, reality is neither at rest nor in motion". (4) is another example of a sentence which turns out false when we read the copula "is" as exemplification and true when read as encoding. Consider (4a):

(4a) $\sim R\Phi_{\bar{E}!} \ \& \sim M\Phi_{\bar{E}!}$.

Since we have defined Platonic Being, or Reality, as $\Phi_{\bar{E}!}$, (4a) captures (4) when "is" is read as "exemplifies". (4a) is false since $\Phi_{\bar{E}!}$ exemplifies being at rest. But consider (4b):

(4b) $\sim \Phi_{\bar{E}!}R \ \& \sim \Phi_{\bar{E}!}M$.

The key to seeing that this might be right comes from the following definition:

$\boldsymbol{D}_9$ *The nature of* $\Phi_F =_{df} F$.

The nature of a Form is the property it encodes. Thus, we read "in virtue of its own nature" as a clue to thinking that Plato is going to conclude something about the fact that $\bar{E}!$ is central to the identity of $\Phi_{\bar{E}!}$.

Assumption A_2 tells us that the nature of $\Phi_{\bar{E}!}$ is distinct from the natures of Φ_R and Φ_M. So (4b) is derivable.

Assertion (4) has always been rather puzzling to me, and I think it is interesting that the distinction between exemplifying and encoding a property has helped us to find a true reading for it.

Is there, after all, some unity to the history of philosophy? Do we have here a PRIMA FACIE link between Plato, Meinong, Mally, and the theory of abstract objects? Maybe further investigations along the above lines will help us to answer these questions.

2. MODELLING THE ROUND SQUARE, ETC.

In our first encounter translating certain theoretical statements of natural language into the language of the theory, we discovered that a few of them containing the copula "is" turned out true when translated using an encoding formula yet turned out false when translated using an exemplification formula. In this section, we look at a class of English sentences which exhibit this feature. These sentences can be recognized by the facts that: (1) they have the form "The $F_1, F_2, \ldots, F_n$ is F_i" ($1 \leq i \leq n$), and (2) there is not (or could not be) an object which jointly exemplifies $F_1, F_2, \ldots, F_n$. Here are some examples:

(1) The set of sets which are not members of themselves is a set of sets which are not members of themselves (The F is F).

(2) The round square is round (The F, G is F).

(3) The existent golden mountain is existent (The F, G, H is F).

These sentences seem to be true A PRIORI. But if we translate the description in (2), for example, as "the object which exemplifies roundness and squareness", then the description would fail to denote. It would then be hard to see how to account for the intuitive truth value of the sentence. And similar remarks apply to (1) and (3).

However, if we translate the description in (2) as "the object which encodes just roundness and squareness", and read the "is" as "encodes", we end up with the truth: the object which encodes just roundness and squareness encodes roundness. In a similar manner, we read (1) as: the object which encodes just being a set of sets which are not self-members encodes being a set of sets which are not self-members. And we do something similar for (3). The suggestion, then, is to translate "The $F_1,\ldots,F_n$ is F_i" as "the object which encodes just $F_1,\ldots,F_n$ encodes F_i".

To make this suggestion precise, we must focus on an interesting class of descriptions. These are descriptions of the form: $(\iota x)(A!x \;\&\; (F)(xF \equiv \chi))$. We call this class of descriptions **A-object descriptions**, and the reason they are interesting is that whenever χ is a formula with no free x's, the resulting description always has a denotation. This is a consequence of the UNIQUENESS theorem schema for objects. In fact, UNIQUENESS and DESCRIPTIONS allow us to prove an interesting set of theorems governing the A-object descriptions:

THEOREM(S) ("A-DESCRIPTIONS"): $\chi_F^G \equiv (\iota x)(A!x \;\&\; (F)(xF \equiv \chi))G$.

Proof. ($\rightarrow$) Suppose G satisfies χ. By UNIQUENESS, there is a unique A-object, say b, which encodes exactly the properties which satisfy χ. So b encodes G. So there is a unique A-object which encodes exactly the properties satisfying χ and something which encodes exactly the properties satisfying χ also encodes G. So by DESCRIPTIONS, the A-object which encodes exactly the properties which satisfy χ encodes G.[11] ($\leftarrow$) By reversing the reasoning. ⊠

Using this theorem schema, it now becomes possible to prove certain facts regarding the objects denoted by A-descriptions. Consider $(\iota x)(A!x \;\&\; (F)(xF \equiv F=R \vee F=S)$, where "$R$" denotes roundness and "S" denotes squareness. If we let $\chi=(F=R \vee F=S)$, then χ_F^R and χ_F^S. So by A-DESCRIPTIONS, $(\iota x)(A!x \;\&\; (F)(xF \equiv F=R \vee F=S))$ encodes both R and S, as we might have expected. In general, when $\chi=(G=F_1 \vee G=F_2 \vee \ldots \vee G=F_n)$, it is provable that:

$$(\iota x)(A!x \;\&\; (G)(xG \equiv G=F_1 \vee \ldots \vee G=F_n))F_i,$$

where $1 \leq i \leq n$, given A-DESCRIPTIONS.

This provides us with the key to the proper translation of our data. For simplicity, let us shorten A-descriptions by using restricted variables to range over A-objects. In fact, throughout the remainder of this work, we use z-variables to range over A-objects.[12] So our A-object descriptions now have the form: $(\iota z)(F)(zF \equiv \chi)$. Where "$S$" denotes being a set, "$\in$" denotes the membership relation, and where the other abbreviations are obvious, we may translate the descriptions in (1)–(3) as (a)–(c), respectively:

(a) $(\iota z)(F)(zF \equiv F = [\lambda x\, Sx \;\&\; (y)(y \in x \equiv Sy \;\&\; y \notin y)])$[13]

(b) $(\iota z)(F)(zF \equiv F = R \vee F = S)$

(c) $(\iota z)(F)(zF \equiv F = E! \vee F = G \vee F = M)$.

And in general, the descriptions in the class of English sentences we have singled out are to be translated as:

$$(\iota z)(G)(zG \equiv G = F_1 \vee G = F_2 \vee \ldots \vee G = F_n).$$

In the metalanguage, we will signal the fact that we intend this reading of the English definite article by writing "the$_{\mathscr{A}}$".

Now let $(\iota z)\psi_1, (\iota z)\psi_2$, and $(\iota z)\psi_3$ abbreviate the descriptions in (a)–(c), respectively. We then translate (1)–(3) into our language as (1)′–(3)′, respectively:

(1)′ $(\iota z)\psi_1[\lambda x\, Sx \;\&\; (y)(y \in x \equiv Sy \;\&\; y \notin y)]$

(2)′ $(\iota z)\psi_2 R$

(3)′ $(\iota z)\psi_3 E!$

(1)′–(3)′ are all theorems, hence the A PRIORI character of the English. In general, our translation of "The $F_1, \ldots, F_n$ is F_i", where there is not (or could not be) an object which jointly exemplifies $F_1, \ldots, F_n$, will always be a theorem of the following form:

$$(\iota z)(G)(zG \equiv G = F_1 \vee \ldots \vee G = F_n)F_i.$$

There is a closely related use of the English definite article. Here are some examples:

(4) The even prime number greater than two is not odd.

(5) The set of all non-self-membered sets is a set.

(6) The existent golden mountain has a shape.

(7) Necessarily, the teacher of Aristotle is a teacher.

These sentences will be represented with the help of a slightly modified A-description. "The even prime number greater than two" shall be translated as "the A-object which encodes being an even prime number greater than two or any property implied by this property". To represent and interpret this reading of the definite article, we must define "F implies

G" as necessarily, everything exemplifying F exemplifies G. So we postpone further investigation until the modal theory has been developed.

3. THE PROBLEM OF EXISTENCE

The property of existence has puzzled philosophers for years. The assertion that some particular thing fails to exemplify existence (or being) strangely carries with it a commitment to the existence (or being) of the very thing which serves as the subject of the assertion. This is partly a result of trying to keep the theory of language as simple as possible – we try to account for the truth of a simple sentence by supposing that the objects denoted by the object terms are in an extension of the relation denoted by the relation term. But when we have a true non-existence claim, talk about "the object denoted by the object name" seems illegitimate.

Although the theory we have developed is rather flexible on this issue, our discussion of the matter will be slightly complicated by the fact that we have taken "existence" as a primitive theoretical notion. The reason this may confuse things is that this primitive notion is not necessarily the notion to use to translate the English word "exists", as it occurs in the data. To see this, consider first the fact that the following two sentences are theorems:

(1) $[\lambda y \sim E!y](\imath z)(F)(zF \equiv F=G \vee F=M)$

(2) $\sim E!(\imath z)(F)(zF \equiv F=G \vee F=M)$.

That is, it is provable both that the$_{\mathscr{A}}$ golden mountain exemplifies non-existence and that it fails to exist. However, neither (1) nor (2) would be an acceptable translation of (3):

(3) The golden mountain does not exist.

(3) has at least one reading on which it is contingent and not knowable A PRIORI. So that eliminates both (1) and (2) as acceptable translations.

However, an abstract object x can "exist" in the sense that some existing object exemplifies all the properties x encodes:

$\boldsymbol{D}_{10}$ x *exists*$_2$("$E!!x$")$=_{df}(\exists y)(E!y \;\&\; (F)(xF \rightarrow Fy))$

It cannot be known A PRIORI that the$_{\mathscr{A}}$ golden mountain fails to exist$_2$. That is, we may read (3) as the contingent (3′):

(3′) $\sim E!!(\iota z)(F)(zF \equiv F=G \vee F=M)$

The English word "exists" is therefore properly translated as "$E!!$". Meinong could have truthfully responded to Russell that the$_{\mathcal{A}}$ existent golden mountain is existent but doesn't exist$_2$.

It is important to realize that we were not forced to have theorems like (1) and (2) above. We could have designed things so that it was provable that everything whatsoever exists. Instead of taking existence as our primitive theoretical relation, we could have started with the notion of being abstract ("$A!$"). We could have then defined:

x is concrete ("$C!x$") $=_{df} [\lambda y \sim A!y]x$

x exists ("$E!x$") $=_{df} A!x \vee C!x$.

We could have then revised NO-CODER as: $C!x \rightarrow \sim(\exists F)xF$. Finally, we could have relabeled "$=_E$" as "$=_C$," changed E-IDENTITY to C-IDENTITY (i.e., $x=_C y \equiv C!x \,\&\, C!y \,\&\, (F)(Fx \equiv Fy)$), and redefined general identity (i.e., $x=y=_{df} x=_C y \vee (A!x \,\&\, A!y \,\&\, (F)(xF \equiv yF))$). Leaving A-OBJECTS as it stands, we could call the result of all these changes VERSION 2.

On VERSION 2, it is provable that everything whatsoever exists. VERSION 2 can do all the work the original theory can do. That is because the exemplification/encoding distinction, and the distinction between two types of objects, remain intact. On VERSION 2, we still have to analyze (3) in a manner analogous to the above. To translate "the golden mountain does not exist" properly, we have to suppose that "the golden mountain" denotes the$_{\mathcal{A}}$ golden mountain and that the sentence claims about this object that there are (exist) no concrete objects which exemplify all the properties it encodes.

So the theory is really pretty flexible on the question are there objects which fail to exist? It is a question of how you prefer to use the word "exists". But I think that philosophers who insist that VERSION 2 is the only correct version of the theory are mistaken. The theory remains useful no matter which of the two versions you adopt. We have taken the present course because it leaves us with a formal language which can be used to investigate the claim that there is a distinction in natural language between the quantifiers "there is" and "there exists". Some philosophers, myself included, believe that there is an exploitable difference in meaing between these two quantifiers of English. Our view can be made precise by investigating a language in which this difference in meaning might be represented. The language we have now is such a language. We use "$(\exists x)\phi$" to express the fact that THERE IS an x such that ϕ, and use "$(\exists x)(E!x \ \& \ \phi)$" to express the fact that THERE EXISTS an x such that ϕ. In a theory which supposed that all the things there are exist, there is no natural way to do this. But this is not an overriding reason for keeping things as they are.

Meinong claimed that "the Object as such stands beyond being and non-being" and that "the Object is by nature indifferent to being".[14] I am not a Meinong scholar, so I do not suppose that I know what Meinong meant by this "doctrine of aussersein", and I do not suppose that he had these two versions of our theory before his mind when he said things like this. Nevertheless, something like these cryptic utterances of Meinong are relevant here. It just does not matter whether you conceive of A-objects as existing or as failing to exist.

Maybe the word "exists" is an ambiguous word, one of the senses of which is a property which has a negation that also turns out to be a sense of the word.[15] To make this idea plausible, we could stick with the original version of the theory, and read "$E!$" as "real existence" and "$[\lambda x \sim E!x]$" as "Platonic existence". Now we have two kinds of existence, with A-objects exemplifying the latter kind. This reading of $[\lambda x \sim E!x]$, besides working to our advantage in Section 1, is further justified by the facts that in the modal theory which follows, A-objects end up having being in every possible world and the class of A-objects stays fixed from world to world. Platonic beings are necessary beings, and A-objects turn out to be necessary beings.[16] They, therefore, exhibit a more perfect kind of existence.[17]

So talking in terms of two kinds of existence is yet a third way of approaching the problem of existence. This means that we really do not have to commit ourselves on the question: Do A-objects fail to exist? Three equivalent versions of the theory decide the question in different ways. The version one prefers to go with will be mostly a result of a decision about which of the various senses of the word "exist" one prefers to use.[18] We shall use it to mean "having a location in space-time."

APPENDIX TO CHAPTER II

In this appendix, we describe the modifications and restrictions which have to be incorporated in the system of Chapter I in order to accommodate descriptions.

(A) The first thing to do is to revise the simultaneous recursive definition of formula and n-place relation term so that the descriptions get generated in a recursive clause defining new object terms. This is relatively straightforward and the result will be very similar to the definition in III., Section 1, B.

(B) Next, we semantically interpret these descriptions by adding a clause to I., Section 2, B., *Denotations*. The clause should read:

$$d_{\mathcal{I},f}((\iota x)\phi) = \begin{cases} o \text{ iff } (\exists f')(f' \overset{x}{=} f \;\&\; f'(x) = o \;\&\; f' \text{ satisfies } \phi \;\&\; \\ \quad (f'')(f'' \overset{x}{=} f \;\&\; f'' \text{ satisfies } \phi \rightarrow f'' = f')) \\ \text{undefined, otherwise.} \end{cases}$$

This guarantees that $(\iota x)\phi$ denotes the unique object o satisfying the description, if there is one, and denotes nothing at all if there is not one. Note that if a description fails to denote, the failure of denotation is inherited by any complex term in which it occurs. So if "$(\iota x)Gx$" does not denote, neither will "$[\lambda y\, Ty(\iota x)Gx]$" nor "$(\iota y)Ty(\iota x)Gx$".

(C) We must next prevent the base clauses in the definition of satisfaction from being undefined. So I., Section 2, C., clauses (1) and (2) need to be redesigned and should read as follows:

(1) If $\phi = \rho^n o_1 \ldots o_n$, f satisfies ϕ iff $(\exists o_1)\ldots(\exists o_n)(\exists r^n)(o_1 = d_{\mathcal{I},f}(o_1) \;\&\ldots\&\; o_n = d_{\mathcal{I},f}(o_n) \;\&\; r^n = d_{\mathcal{I},f}(\rho^n) \;\&\; \langle o_1, \ldots, o_n \rangle \in ext_{\mathcal{R}}(r^n))$.

(2) If $\phi = o\rho^1$, f satisfies ϕ iff $(\exists o)(\exists r^1)(o = d_{\mathcal{I},f}(o) \;\&\; r^1 = d_{\mathcal{I},f}(\rho^1) \;\&\; o \in ext_{\mathcal{A}}(r^1))$.

So if an assignment f is to satisfy an atomic formula, all the terms in the formula must have a denotation. The other clauses in the definition of satisfaction are acceptable as stated.

(D) Fourthly, we must modify one of our logical axioms so that we may invalidate the following proof of a proper theorem which would be false in some models of the theory:

(i)	$(x)(x = x)$	=I(Proper Theorem)
(ii)	$(\iota x)Gx = (\iota x)Gx$	LA4
(iii)	$(\exists y)(y = (\iota x)Gx)$	EI.

Existential Introduction ("EI") is a typical derived rule of inference and can be used to move from line (ii) to line (iii).[19] Although line (iii) abbreviates a longer formula, it will be false in models of the theory in which $(\iota x)Gx$ fails to denote. The problem here is that LA4, as it now stands, allows us to instantiate universals to terms which might fail to denote. This gets us into trouble when a true universal claim gets converted, by LA4, into a formula in which atomic formulas appear containing the non-denoting term.

Let us say that a term τ **contains a definite description** iff either τ is a definite description or a definite description occurs somewhere in τ. For example, the following three terms all contain descriptions: $(\iota x)Gx$, $(\iota x)Tx(\iota y)Hy$, $[\lambda y\, Ty(\iota x)Gx]$. Now the only terms of our language that might fail to denote in an arbitrary interpretation are those which contain descriptions. All primitive names are guaranteed a denotation, since the domains of interpretation must be non-empty and the $\mathscr{F}$ function is a total function from the set of primitive names into the appropriate domains. Also, any λ-expression which doesn't contain a description is guaranteed a denotation. The λ-expressions are partitioned and to each equivalence class there corresponds a unique clause in the definition of denotation$_{\mathscr{I},f}$. So if each term in the λ-expression has a denotation, the logical functions in $\mathscr{L}$ guarantee that a relation with the appropriate structure will be found in the domain of relations.

LA4 will never get us into trouble therefore if we require that only terms which are guaranteed denotations (i.e. terms which do not contain descriptions) may instantiate universal claims. But we also want to instantiate universal claims to terms which contain descriptions whenever we know such terms have a denotation. This happens whenever such terms appear in true atomic formulas. So if a universal claim is true, then if a term containing a description appears in a true atomic formula, the result of instantiating that term into the universal claim should also be true.

These remarks can be incorporated into our system by revising LA4 into the following two axiom schemata:

LA4a: $(\alpha)\phi \rightarrow \phi^{\tau}_{\alpha}$, where τ is substitutable for α and τ contains no descriptions

LA4b: $(\alpha)\phi \rightarrow (\psi^{\tau}_{\beta} \rightarrow \phi^{\tau}_{\alpha})$, where τ is substitutable for both α and β, and ψ is any atomic formula.

The reader may wonder here whether we have considered the option standardly taken in predicate calculi containing descriptions. In a typical second order predicate calculus where identity formulas are primitive and interpreted in the usual way, where the domains of interpretation must be non-empty, where the usual two identity axioms have been added to the logic, and where the only terms which might fail to denote are those which contain descriptions, the normal way of invalidating inferences like the above is to revise LA4 to the following two schemata:

$(\alpha)\phi \rightarrow ((\exists\beta)\beta = \tau \rightarrow \phi^{\tau}_{\alpha})$, where τ is substitutable for α
$(\exists\beta)\beta = \tau$, where τ contains no descriptions.

Instances of the first would be logically true in our system, strangely enough, despite the fact that the identity symbol is defined in various ways. But some instances of the second would not be logically true. Consider instances in which the quantified variable is an object variable, for example, $(\exists x)x = c$. This abbreviates $(\exists x)(x =_E c \vee (A!x \;\&\; A!c \;\&\; (F)(xF \equiv cF)))$. Since $=_E$ denotes a primitive relation of the theory, consider interpretations in which the proper axiom *E*-IDENTITY is false. For example, consider an interpretation in which $=_E$ denotes an irreflexive relation and *c* denotes an existing object. Why should $(\exists x)x =_E c$ be true in that interpretation?

Consequently, this revision of LA4 will not help us, and we shall adopt LA4a and LA4b as the official logical axioms of our system. It should be easy to see that they are both logically true. LA4a blocks the undesirable inferences in our system, while LA4b allows us to instantiate universal claims with terms containing denoting descriptions.[20] Of course we can always "instantiate" definitions with descriptions, even if the descriptions fail to denote. Definitions are not universal claims. They are metalinguistic conventions for abbreviations. Strictly speaking, only metavariables ranging over the appropriate terms should be used in introducing the definitions. However, we employ object language variables for convenience, since it makes it easier to read the formulas.

(E) The next modification we need to make in order to successfully incorporate descriptions concerns our λ-EQUIVALENCE schema. We must require that none of the λ-expressions used in the schema are constructed out of propositional formulas ϕ in which there occur descriptions. This prevents the following derivation of a formula which is not valid:

(i) $(x)([\lambda y\, Fy(\iota u)Gu \vee \sim Fy(\iota u)Gu]x \equiv Fx(\iota u)Gu \vee \sim Fx(\iota u)Gu)$
by λ-EQUIVALENCE

(ii) $[\lambda y\, Fy(\iota u)Gu \vee \sim Fy(\iota u)Gu]a \equiv Fa(\iota u)Gu \vee \sim Fa(\iota u)Gu$
by LA4a.

(ii) will be false in interpretations where "$(\iota u)Gu$" fails to denote, since the right side of the biconditional would be true while the left side false. That is because the failure of the description to denote is inherited by the λ-expression. But for the atomic formula constituting the left side of the biconditional to be true, both the object term and the complex relation term must have a denotation.

Consequently, λ-EQUIVALENCE should be reformulated as follows:

λ-EQUIVALENCE: where ϕ is any propositional formula with no descriptions, the following is an axiom:

$$(x_1)\ldots(x_n)([\lambda v_1 \ldots v_n\, \phi]x_1 \ldots x_n \equiv \phi^{x_1,\ldots,x_n}_{v_1,\ldots,v_n}).$$

Then, by using LA4b, we may construct the following derivation:

(i) $(u)(x)([\lambda y\, Fyu \vee \sim Fyu]x \equiv Fyu \vee \sim Fyu)$
λ-EQUIVALENCE and UI

(ii) $\psi^{(\iota u)Gu}_{v} \rightarrow ([\lambda y\, Fy(\iota u)Gu \vee \sim Fy(\iota u)Gu]a \equiv Fa(\iota u)Gu \vee$
$\sim Fa(\iota u)Gu)$ by UE and LA4b, where ψ is atomic.

(F) Next, we investigate the logical axioms which must be added to the logic if we are to be able to derive logical truths which arise specifically as a result of the semantic interpretation of descriptions. The question is complicated by the following two facts: (1) that identity is a primitive logical notion in the semantics and crucially appears in the clause assigning a denotation to descriptions (see part B, above), and (2) that identity is not a primitive logical notion of the object language, and as we have it defined, it works like it should only in the presence of the proper axioms E-IDENTITY and IDENTITY. Were it not for these facts, we could have just added an axiom like DESCRIPTIONS and be done with it. But, as noted in note 1, DESCRIPTIONS is not logically true. And for the reasons mentioned in that note, I do not think that any other formulas utilizing defined identity which might prove useful for our logic are going to be logically true. So which formulas without identity are the logical truths specifically relating to descriptions from which all others can be derived?

Clearly the following two formulas are both logically true: $H(\iota x)Gx \rightarrow (\exists y)(Gy \,\&\, Hy)$ and $(H(\iota x)Gx \,\&\, F(\iota x)Gx) \rightarrow (\exists y)(Gy \,\&\, Hy \,\&\, Fy)$. So we need to add the following general schema:[21]

L-DESCRIPTIONS$_1$: where ψ is any atomic formula or conjunction of atomic formulas, the following is an axiom:

$$\psi_v^{(\iota x)\phi} \rightarrow (\exists y)(\phi_x^y \,\&\, \psi_v^y).$$

But this schema, by itself, does not seem to be sufficient. We do not seem to be able to derive $H(\iota x)Gx \rightarrow (y)(Gy \rightarrow Hy)$. So we probably need to add:

L-DESCRIPTIONS$_2$: where ψ is any atomic formula, the following is an axiom:

$$\psi_v^{(\iota x)\phi} \rightarrow (y)(\phi_x^y \rightarrow \psi_v^y).$$

But even this does not seem to be sufficient, since it does not look like we will be able to derive $H(\iota x)Gx \rightarrow \sim((\exists y)(Gy \,\&\, Fy) \,\&\, (\exists y)(Gy \,\&\, \sim Fy))$. Again, it looks like the following schema should be added:

L-DESCRIPTIONS$_3$: where ψ is any atomic formula with v_1 free and χ is any formula with v_2 free, the following is an axiom:

$$\psi_{v_1}^{(\iota x)\phi} \rightarrow \sim((\exists y)(\phi_x^y \,\&\, \chi_{v_2}^y) \,\&\, (\exists y)(\phi_x^y \,\&\, \sim \chi_{v_2}^y)).$$

Of course, only a completeness proof will verify that these three logical axioms will be sufficient. I suspect, however, that there are still underivable logical truths involving descriptions. Given our present concern with metaphysics, I think that we may feel secure that our proper axiom DESCRIPTIONS will allow us to prove all of the consequences of the theory which we will need in the applications.

Logicians should be interested in the project of developing a complete logic for descriptions in a language without identity. They may end up with a rather inelegant group of logical axioms. But that would simply be because the expressive power of the object language falls far short of the expressive power of the language used in the semantics, which takes identity as an extra primitive. But I prefer keeping the object language elegant and complicating the logic to complicating the object language with the addition of primitive identity. We must certainly be justified in searching for the smallest set of primitives powerful enough to derive a given set of interesting results. Identity is a concept which we can ANALYZE in terms of our other primitives. I take the definitions of identity, $\boldsymbol{D}_2$, $\boldsymbol{D}_3$, and $\boldsymbol{D}_4$ to be insightful. We no longer have to wonder what philosophers and model-theoretic logicians mean when they appeal to this notion. And if in the course of trying to define as much as possible in terms of a few potent primitives we have to add several logical axioms to guarantee completeness, so be it (there are also other reasons for not adding identity as a primitive to the object language, for example, we would have to place restrictions on λ-formation and λ-EQUIVALENCE, to avoid McMichael's paradox (see Appendix A, part A)).

(G) We close the appendix with a few remarks on the proper axiom DESCRIPTIONS. Note that it allows us to prove the important set of proper theorems: $\psi_v^{(\iota x)\phi} \rightarrow (\exists y)(y=(\iota x)\phi)$, where ψ is any atomic formula or any defined identity formula with one free object variable v. For if $\psi_v^{(\iota x)\phi}$, then by DESCRIPTIONS, $(\exists! y)\phi_x^y$. If we call an arbitrary such object b, then we know $\phi_x^b \,\&\, (u)(\phi_x^u \rightarrow u=b)$. Now if we can show $b=(\iota x)\phi$, then since there are atomic formulas in the defined notation, we may use EI to reap our result.[22] We prove $b=(\iota x)\phi$ by using DESCRIPTIONS, this time deriving the right side of the biconditional with ψ as an identity formula. That is, we try to derive $(\exists! y)\phi_x^y \,\&\, (\exists y)(\phi_x^y \,\&\, b=y)$. We already have the first conjunct. The second is easily obtained given that we know ϕ_x^b and that $b=b$ is introducable by $=$I. So $(\exists y)(\phi_x^y \,\&\, b=y)$. So by DESCRIPTIONS, $b=(\iota x)\phi$, and by EI, $(\exists y)(y=(\iota x)\phi)$.

DESCRIPTIONS forces all of the models of the theory to be such that objects which satisfy $\boldsymbol{D}_4$ (Chapter I, Section 4) are *identical*.[23] To see this, consider the following instance: $H(\iota x)Gx \equiv (\exists! y)Gy \,\&\, (\exists y)(Gy \,\&\, Hy)$. Whenever the right side is true, some object, say *o*, exemplifies G and everything else which exemplifies G is either E-identical with *o* or is abstract and encodes the properties *o* encodes. Semantically, there may be some *distinct* object other than *o* which in fact does this. But in all the models of the theory, the left side of DESCRIPTIONS will then force that object to be *identical* to *o*, given the way we have interpreted descriptions in part B above.

The above remarks on incorporating descriptions should give the reader a fairly good idea of how our system adjusts to the acquisition of terms which may fail to denote.

CHAPTER III

THE MODAL THEORY OF ABSTRACT OBJECTS (WITH PROPOSITIONS)

1. THE LANGUAGE

A. PRIMITIVE SYMBOLS

To the language of Chapter I, we add the "□"-operator (to express the English sentential adverb "necessarily") and names of (and variables ranging over) propositions. By allowing the superscripts on the primitive relation terms to reach zero, we obtain names and variables for propositions. For convenience, we use $P^0, Q^0, R^0, \ldots$ and $F^0, G^0, H^0, \ldots$ as names and variables, respectively, for propositions. Officially, however, our new list of primitive symbols is as follows:

(1) Primitive object terms
Names: $a_1, a_2, \ldots$
Variables: $x_1, x_2, \ldots$

(2) Primitive n-place relation terms
Names: $P_1^n, P_2^n, \ldots, =_E, E!$ $\quad n \geq 0$
Variables: $F_1^n, F_2^n, \ldots$

(3) Connectives: $\sim$, $\rightarrow$

(4) Quantifier: $\forall$

(5) Lambda: λ

(6) Iota: ι

(7) Box: □

(8) Parentheses and brackets: (,), [,].

B. FORMULAS AND TERMS

We simultaneously define **(propositional) formula**, **object term**, and **n-place relation term**, inductively, as follows:

(1) All primitive object terms are object terms and all primitive n-place relation terms are n-place relation terms.

(2) If ρ^0 is any zero-place relation term, ρ^0 is a (propositional) formula.

(3) Atomic exemplification: If ρ^n is any n-place relation term and $o_1, \ldots, o_n$ are any object terms, $\rho^n o_1 \ldots o_n$ is a (propositional) formula.

(4) Atomic encoding: If ρ^1 is any one-place relation term, and o is any object term, $o\rho^1$ is a formula.

(5) Molecular, Quantified, and Modal: If ϕ and ψ are any (propositional) formulas, and α is any (object) variable, then $(\sim \phi)$, $(\phi \rightarrow \psi)$, $(\forall\alpha)\phi$, and $(\Box\, \phi)$ are (propositional) formulas.

(6) Object descriptions: If ϕ is any formula with one free object variable x, then $(\iota x)\phi$ is an object term.

(7) Complex n-place relation terms: If ϕ is any propositional formula, and $v_1, \ldots, v_n$ are any object variables which may or may not be free in ϕ, then $[\lambda v_1, \ldots, v_n \phi]$ is an n-place relation term $(n \geq 1)$ and ϕ itself is a zero-place relation term (in what follows, it will sometimes be convenient to regard ϕ as a degenerate λ-expression, $[\lambda\phi]$, when $n = 0$).

In addition to the standard abbreviations for the connectives and quantifiers, we use $\Diamond\phi$ to abbreviate $\sim\Box\sim\phi$. However, we now define:

D_1 *x is abstract* ("$A!x$") $=_{df} [\lambda y \Box \sim E!y]x$

D_2 *x is a possibly existing object* $=_{df} \Diamond E!x$.

So abstract objects are just not the kind of thing that could exist. Here are some examples of schemata and formulas: $\Box Q$ ("it is necessary that Q"); $\Box(\exists x)(A!x\ \&\ (F)(xF \equiv \phi))$ ("necessarily, some abstract object encodes exactly the properties satisfying ϕ"); $\Diamond(\exists y)(F)(xF \rightarrow Fy)$ ("possibly, there is an object which exemplifies every property x encodes"); and $(x)(\Diamond E!x \rightarrow \sim(\exists F)xF)$ ("possibly existing objects fail to encode any properties").

We say that a formula ϕ **necessarily implies** a formula ψ ("$\phi \Rightarrow \psi$") iff $\Box(\phi \rightarrow \psi)$. ϕ is **necessarily equivalent** to ψ ("$\phi \Leftrightarrow \psi$") iff $\Box(\phi \equiv \psi)$.

There are two kinds of complex terms – object descriptions and complex n-place relation terms. Modal formulas may appear in both. For example, $(\iota x)(A!x\ \&\ (F)(xF \equiv F = R))$ is an object description which reads: the abstract object which encodes just R.[1] The inductive clause for complex n-place relation terms differs from its counterpart in the elementary theory in three important respects: (1) it allows modal formulas to appear after

λ's if the formula is propositional, (2) it allows λ's to bind variables which are not free in the ensuing formula, and (3) it allows propositional formulas themselves to be relation terms. Here are some examples of new complex n-place relation terms: $[\lambda xy \Box Qb]$ ("being a first thing and a second thing such that necessarily, b exemplifies Q"); $[\lambda x \Box(E!x \rightarrow Px)]$ ("being an x such that necessarily, if x exists, x exemplifies P"); $\Box Gb$ ("b exemplifies G essentially").

As before, τ is a **term** iff either τ is an object term or τ is an n-place relation term, for some n.

2. THE SEMANTICS

A. INTERPRETATIONS

An interpretation, $\mathscr{I}$, of our modified second order modal language is any octuple, $\langle \mathscr{W}, w_0, \mathscr{D}, \mathscr{R}, ext_w, \mathscr{L}, ext_{\mathscr{A}}, \mathscr{F} \rangle$, which meets the conditions described in this subsection. The first member of $\mathscr{I}$ is a non-empty class, $\mathscr{W}$, called the class of **possible worlds.**[2] The second member of $\mathscr{I}$, w_0, is chosen from $\mathscr{W}$ and is called the **actual world**. The third member, $\mathscr{D}$, is a non-empty class and is called the **domain of objects**. The fourth member, $\mathscr{R}$, is also a non-empty class, and is called the **domain of relations**. $\mathscr{R}$ is the union of a sequence of non-empty classes $\mathscr{R}_0, \mathscr{R}_1, \mathscr{R}_2, \ldots$, i.e., $\mathscr{R} = \bigcup_{n \geq 0} \mathscr{R}_n$. Each $\mathscr{R}_n$ is called the class of **n-place relations** (we call $\mathscr{R}_1$ the class of **properties**, and $\mathscr{R}_0$ the class of **propositions**). $\mathscr{R}$ must be closed under all the logical functions specified in the sixth member of the interpretation ($\mathscr{L}$).

The fifth, sixth, and seventh members of $\mathscr{I}$ impose a structure on $\mathscr{W}$, $\mathscr{D}$, and $\mathscr{R}$. The fifth member of $\mathscr{I}$, ext_w, is a function which maps $\mathscr{R}_n \times \mathscr{W}$ into $\mathfrak{P}(\mathscr{D}^n)$ ("the power set of $\mathscr{D}^n$"), where $n \geq 1$, and which maps $\mathscr{R}_0 \times \mathscr{W}$ into $\{T, F\}$. We index the function to its second argument and call $ext_w(\imath^n)$ the **exemplification extension** ("extension$_w$") of $\imath^n$ **at** w.

The sixth member of $\mathscr{I}$, $\mathscr{L}$, is a class of logical functions which operate in a manner similar to their counterparts in the semantics of the elementary theory. However, we: (1) add two additional functions, $\mathscr{VAC}_i$ ("i-vacuous expansion") and $\mathscr{NEC}$ ("necessitation"), (2) constrain the extensions$_w$ of the complex relations resulting from all the logical functions at every possible world, and (3) allow $\mathscr{PLUG}_i$ and $\mathscr{UNIV}_i$ to operate on properties, allow $\mathscr{COND}$ and $\mathscr{NEG}$ to operate on propositions, and allow $\mathscr{VAC}_i$ and $\mathscr{NEC}$ to operate on all relations. The definitions which make these three major changes precise go as follows:

(a) $\mathcal{PLUG}_i$, for each i, $i \geq 1$, is a function mapping $(\mathcal{R}_i \cup \mathcal{R}_{i+1} \cup \ldots) \times \mathcal{D}$ into $(\mathcal{R}_{i-1} \cup \mathcal{R}_i \cup \ldots)$ subject to the conditions:

(1) for $n > 1$, $ext_\omega(\mathcal{PLUG}_i(\imath^n, o)) = \{\langle o_1, \ldots, o_{i-1}, o_{i+1}, \ldots, o_n \rangle \mid \langle o_1, \ldots, o_{i-1}, o, o_{i+1}, \ldots, o_n \rangle \in ext_\omega(\imath^n)\}$

(2) for $n = 1$, $ext_\omega(\mathcal{PLUG}_1(\imath^1, o)) = \begin{cases} T \text{ iff } o \in ext_\omega(\imath^1) \\ F \text{ otherwise.} \end{cases}$

(b) $\mathcal{UNIV}_i$, for each i, $i \geq 1$, is a function mapping $(\mathcal{R}_i \cup \mathcal{R}_{i+1} \cup \ldots)$ into $(\mathcal{R}_{i-1} \cup \mathcal{R}_i \cup \ldots)$ subject to the conditions:

(1) for $n > 1$, $ext_\omega(\mathcal{UNIV}_i(\imath^n)) = \{\langle o_1 \ldots, o_{i-1}, o_{i+1}, \ldots, o_n \rangle \mid (\forall o)(\langle o_1, \ldots, o_{i-1}, o, o_{i+1}, \ldots, o_n \rangle \in ext_\omega(\imath^n))\}$

(2) for $n = 1$, $ext_\omega(\mathcal{UNIV}_1(\imath^1) = \begin{cases} T \text{ iff } (\forall o)(o \in ext_\omega(\imath^1)) \\ F \text{ otherwise.} \end{cases}$

(c) $\mathcal{CONV}_{i,j}$, for each i,j, $1 \leq i < j$, is a function mapping $(\mathcal{R}_j \cup \mathcal{R}_{j+1} \cup \ldots)$ into $(\mathcal{R}_j \cup \mathcal{R}_{j+1} \cup \ldots)$ subject to the condition:

$ext_\omega(\mathcal{CONV}_{i,j}(\imath^n)) = \{\langle o_1, \ldots, o_{i-1}, o_j, o_{i+1}, \ldots, o_{j-1}, o_i, o_{j+1}, \ldots, o_n \rangle \mid \langle o_1, \ldots, o_1, \ldots, o_j, \ldots, o_n \rangle \in ext_\omega(\imath^n)\}$.

(d) $\mathcal{REFL}_{i,j}$, for each i,j, $1 \leq i < j$, is a function mapping $(\mathcal{R}_j \cup \mathcal{R}_{j+1} \cup \ldots)$ into $(\mathcal{R}_{j-1} \cup \mathcal{R}_j \cup \ldots)$ subject to the condition:

$ext_\omega(\mathcal{REFL}_{i,j}(\imath^n)) = \{\langle o_1, \ldots, o_i, \ldots, o_{j-1}, o_{j+1}, \ldots, o_n \rangle \mid \langle o_1, \ldots, o_i, \ldots, o_j, \ldots, o_n \rangle \in ext_\omega(\imath^n)$ and $o_i = o_j\}$.

(e) $\mathcal{VAC}_i$, for each i, $i \geq 1$, is a function mapping $(\mathcal{R}_{i-1} \cup \mathcal{R}_i \cup \ldots)$ into $(\mathcal{R}_i \cup \mathcal{R}_{i+1} \cup \ldots)$ subject to the conditions:

(1) for $n \geq 1$, $ext_\omega(\mathcal{VAC}_i(\imath^n)) = \{\langle o_1, \ldots, o_{i-1}, o, o_i, o_{i+1}, \ldots, o_n \rangle \mid \langle o_1, \ldots, o_i, \ldots, o_n \rangle \in ext_\omega(\imath^n)\}$

(2) for $n = 0$, $ext_\omega(\mathcal{VAC}_1(\imath^0)) = \{o \mid ext_\omega(\imath^0) = T\}$

(f) $\mathscr{COND}$ is a function mapping $(\mathscr{R}_0 \cup \mathscr{R}_1 \cup \ldots) \times (\mathscr{R}_0 \cup \mathscr{R}_1 \cup \ldots)$ into $(\mathscr{R}_0 \cup \mathscr{R}_1 \cup \ldots)$ subject to the following conditions:

(1) for $n \geq 1$, $m \geq 1$, $ext_w(\mathscr{COND}(\imath^n, \jmath^m)) = \{\langle o_1, \ldots, o_n, o'_1, \ldots, o'_m\rangle | \langle o_1, \ldots, o_n\rangle \notin ext_w(\imath^n)$ or $\langle o'_1, \ldots, o'_m\rangle \in ext_w(\jmath^m)\}$

(2) for $n = 0$, $m \geq 1$, $ext_w(\mathscr{COND}(\imath^0, \jmath^m)) = \{\langle o_1, \ldots, o_m\rangle | ext_w(\imath^0) = F$ or $\langle o_1, \ldots, o_m\rangle \in ext_w(\jmath^m)\}$

(3) for $n \geq 1$, $m = 0$, $ext_w(\mathscr{COND}(\imath^n, \jmath^0)) = \{\langle o_1, \ldots, o_m\rangle | \langle o_1, \ldots, o_n\rangle \notin ext_w(\imath^n)$ or $ext_w(\jmath^0) = T\}$

(4) for $n = 0$, $m = 0$, $ext_w(\mathscr{COND}(\imath^0, \jmath^0)) =$

$$\begin{cases} T \text{ iff } ext_w(\imath^0) = F \text{ or } ext_w(\jmath^0) = T \\ F \text{ otherwise.} \end{cases}$$

(g) $\mathscr{NEG}$ is a function mapping $(\mathscr{R}_0 \cup \mathscr{R}_1 \cup \ldots)$ into $(\mathscr{R}_0 \cup \mathscr{R}_1 \cup \ldots)$ subject to the conditions:

(1) for $n \geq 1$, $ext_w(\mathscr{NEG}(\imath^n)) = \{\langle o_1, \ldots, o_n\rangle | \langle o_1, \ldots, o_n\rangle \notin ext_w(\imath^n)\}$

(2) for $n = 0$, $ext_w(\mathscr{NEG}(\imath^0)) = \begin{cases} T \text{ iff } ext_w(\imath^0) = F \\ F \text{ otherwise.} \end{cases}$

(h) $\mathscr{NEC}$ is a function mapping $(\mathscr{R}_0 \cup \mathscr{R}_1 \cup \ldots)$ into $(\mathscr{R}_0 \cup \mathscr{R}_1 \cup \ldots)$ subject to the conditions:

(1) for $n \geq 1$, $ext_w(\mathscr{NEC}(\imath^n)) = \{\langle o_1, \ldots, o_n\rangle | (w')(\langle o_1, \ldots, o_n\rangle \in ext_{w'}(\imath^n))\}$

(2) for $n = 0$, $ext_w(\mathscr{NEC}(\imath^0)) = \begin{cases} T \text{ iff } (w')(ext_{w'}(\imath^0) = T) \\ F \text{ otherwise.} \end{cases}$

This completes the definitions of the logical functions.[3] The seventh member of $\mathscr{I}$, $ext_{\mathscr{A}}$, is a function which maps $\mathscr{R}_1$ into $\mathfrak{P}(\mathscr{D})$. $ext_{\mathscr{A}}(\imath^1)$ is called the **encoding extension** ("extension$_{\mathscr{A}}$" in the metalanguage) of $\imath^1$.

The final member of $\mathscr{I}$, the $\mathscr{F}$ function, maps the simple names of the language to elements of the appropriate domain. For each object name κ, $\mathscr{F}(\kappa) \in \mathscr{D}$. For each relation name κ^n, $\mathscr{F}(\kappa^n) \in \mathscr{R}_n$. We call $ext_w(\mathscr{F}(E!))$ the set of **existing** objects **at** w("$\mathscr{E}_w$"). We call $ext_{w_0}(\mathscr{F}(E!))$ the set of

existing objects (i.e., $\mathscr{E} =_{df} \mathscr{E}_{w_0}$). We call $\{o \mid (\exists w)(o \in ext_w(\mathscr{F}(E!)))\}$ the set of **possibly existing** objects ("$\mathscr{PE}$"). The complement of $\mathscr{PE}$ on $\mathscr{D}$ is called the set of **abstract** objects ("$\mathscr{A}$").

B. ASSIGNMENTS AND DENOTATIONS

Partitioning the λ-expressions. Since we have λ-expressions in the modal language which were not part of the elementary language, we must incorporate rules to classify the new possibilities. These new rules correspond to $\mathscr{VAC}_i$ and $\mathscr{NEC}$ – they help to classify λ-expressions with vacuously bound λ-variables and with $\Box$'s.

The following six major rules partition the class of λ-expressions into nine equivalence classes. If μ is an arbitrary λ-expression, $[\lambda v_1 \dots v_n \phi]$, μ is classified as follows:

(1) If $(\exists i)$ ($1 \leq i \leq n$ and v_i does not occur free in ϕ and i is the least such number), then μ is the **i^{th}-vacuous expansion of** $[\lambda v_1 \dots v_{i-1} v_{i+1} \dots v_n \phi]$.

(2) If μ is not an ith-vacuous expansion, then if $(\exists i)$ ($1 \leq i \leq n$ and v_i is not the ith free object variable in ϕ and i is the least such number), then where v_j is the ith free object variable in ϕ, μ is the **i,j^{th}-conversion of** $[\lambda v_1 \dots v_{i-1} v_j v_{i+1} \dots v_{j-1} v_i v_{j+1} \dots v_n \phi]$.

(3) If μ is neither of the above, then

(a) If $\phi = (\sim \psi)$, μ is the **negation of** $[\lambda v_1 \dots v_n \psi]$

(b) If $\phi = (\psi \rightarrow \chi)$, and ψ and χ have no free object variables in common, then where $v_1, \dots, v_p$ are the variables in ψ and $v_{p+1}, \dots, v_n$ are the variables in χ, μ is the **conditionalization of** $[\lambda v_1 \dots v_p \psi]$ **and** $[\lambda v_{p+1} \dots v_n \chi]$

(c) If $\phi = (\forall v)\psi$ and v is the ith free object variable in ϕ, then μ is the **i^{th}-universalization of** $[\lambda v_1 \dots v_{i-1} v v_i v_{i+1} \dots v_n \psi]$

(d) If $\phi = (\Box \psi)$, then μ is the **necessitation of** $[\lambda v_1 \dots v_n \psi]$.

(4) If μ is none of the above, then if $(\exists i)$ ($1 \leq i \leq n$ and v_i occurs free in more than one place in ϕ and i is the least such number), then where:

(a) k is the number of free object variables between the first and second occurrences of v_i,

(b) ϕ' is the result of replacing the second occurrence of v_i with a new variable v, and

(c) $j = i + k + 1$,

μ is the **i, jth-reflection of** $[\lambda v_1 \dots v_{i+k} v v_j \dots v_n \phi']$.

(5) If μ is none of the above, then if o is the left most object term occurring in ϕ, then where:

(a) j is the number of free variables occurring before o,

(b) ϕ' is the result of replacing the first occurrence of o by a new variable v, and

(c) $i = j + 1$,

μ is the **ith-plugging of** $[\lambda v_1 \dots v_j v v_{j+1} \dots v_n \phi']$ **by** o.

(6) If μ is none of the above, then

(a) ϕ is atomic

(b) $v_1, \dots, v_n$ is the order in which these variables first occur in ϕ.

(c) $\mu = [\lambda v_1 \dots v_n \rho^n v_1 \dots v_n]$, for some relation term ρ^n, and

(d) μ is called **elementary**.

In addition to the examples we saw from the elementary theory, we now have: $[\lambda x \Box (Px \rightarrow \sim Qx)]$ is the necessitation of $[\lambda x\, Px \rightarrow \sim Qx]$; $[\lambda xwy\, Px \rightarrow Qy]$ is the 2nd-vacuous expansion of $[\lambda xy\, Px \rightarrow Qy]$; $[\lambda xvwy\, Px \rightarrow Qy]$ is the 2nd-vacuous expansion of $[\lambda xwy\, Px \rightarrow Qy]$; etc.

$\mathscr{I}$-assignments. If given an interpretation $\mathscr{I}$ of our language, an $\mathscr{I}$-assignment, f, will be any function defined on the primitive variables of the language satisfying the following two conditions:

(1) where v is any object variable, $f(v) \in \mathscr{D}$

(2) where π^n is any relation variable, $f(\pi^n) \in \mathscr{R}_n$.

Denotations. If given an interpretation $\mathscr{I}$ of our language, and an $\mathscr{I}$-assignment f, then we recursively define **the denotation of** term τ **with respect to** interpretation $\mathscr{I}$ **and** $\mathscr{I}$-assignment f ("$d_{\mathscr{I},f}(\tau)$") as follows:

(1) where κ is any primitive name, $d_{\mathscr{I},f}(\kappa) = \mathscr{F}(\kappa)$

(2) where α is any primitive variable, $d_{\mathscr{I},f}(\alpha) = f(\alpha)$

(3) where $(\iota x)\phi$ is any object description,

$$d_{\mathscr{I},f}((\iota x)\phi) = \begin{cases} o \text{ iff } (\exists f')(f' \overset{*}{_x} f \;\&\; f'(x) = o \;\&\; f' \text{ satisfies } \phi \text{ with respect to } w_0 \;\&\; (f'')(f'' \overset{*}{_x} f \;\&\; f'' \text{satisfies } \phi \text{ with respect to } w_0 \rightarrow f'' = f')) \\ \text{undefined, otherwise.} \end{cases}$$

where satisfaction is defined as in subsection C.[5]

(4) where $[\lambda v_1 \ldots v_n \rho^n v_1 \ldots v_n]$ is any elementary λ-expression,

$d_{\mathscr{I},f}([\lambda v_1 \ldots v_n \rho^n v_1 \ldots v_n]) = d_{\mathscr{I},f}(\rho^n)$

(5) where μ is the ith-plugging of ξ by o,

$d_{\mathscr{I},f}(\mu) = \mathscr{PLUG}_i(d_{\mathscr{I},f}(\xi), d_{\mathscr{I},f}(o))$

(6) where μ is the ith-universalization of ξ,

$d_{\mathscr{I},f}(\mu) = \mathscr{UNIV}_i(d_{\mathscr{I},f}(\xi))$

(7) where μ is the i,j^{th}-conversion of ξ,

$d_{\mathscr{I},f}(\mu) = \mathscr{CONV}_{i,j}(d_{\mathscr{I},f}(\xi))$

(8) where μ is the i,j^{th}-reflection of ξ,

$d_{\mathscr{I},f}(\mu) = \mathscr{REFL}_{i,j}(d_{\mathscr{I},f}(\xi))$

(9) where μ is the ith-vacuous expansion of ξ,

$d_{\mathscr{I},f}(\mu) = \mathscr{VAC}_i(d_{\mathscr{I},f}(\xi))$

(10) where μ is the conditionalization of ξ and ζ,

$d_{\mathscr{I},f}(\mu) = \mathscr{COND}(d_{\mathscr{I},f}(\xi), d_{\mathscr{I},f}(\zeta))$

(11) where μ is the negation of ξ,

$d_{\mathscr{I},f}(\mu) = \mathscr{NEG}(d_{\mathscr{I},f}(\xi))$

(12) where μ is the necessitation of ξ,

$d_{\mathscr{I},f}(\mu) = \mathscr{NEC}(d_{\mathscr{I},f}(\xi))$

(13) where ϕ is any propositional formula, $d_{\mathscr{I},f}(\phi)$ is defined as follows:

(a) if ϕ is a primitive zero-place term, $d_{\mathscr{I},f}(\phi)$ is defined above

(b) if $\phi = \rho^n o_1 \ldots o_n, d_{\mathscr{I},f}(\phi) =$
$\mathscr{PLUG}_1(\mathscr{PLUG}_2(\ldots(\mathscr{PLUG}_n(d_{\mathscr{I},f}(\rho^n), d_{\mathscr{I},f}(o_n)), \ldots),$
$d_{\mathscr{I},f}(o_2)), d_{\mathscr{I},f}(o_1))$

(c) if $\phi = (\sim\psi), d_{\mathscr{I},f}(\phi) = \mathscr{NEG}(d_{\mathscr{I},f}(\psi))$

(d) if $\phi = (\psi \to \chi),\ d_{\mathscr{I},f}(\phi) = \mathscr{COND}(d_{\mathscr{I},f}(\psi), d_{\mathscr{I},f}(\chi))$

(e) if $\phi = (\forall v)\phi, d_{\mathscr{I},f}(\phi) = \mathscr{UNIV}_1(d_{\mathscr{I},f}([\lambda v \psi]))$

(f) if $\phi = (\Box\psi), d_{\mathscr{I},f}(\phi) = \mathscr{NEC}(d_{\mathscr{I},f}(\psi))$.

Here are some examples of λ-expressions and their denotations with respect to a given $\mathscr{I}$ and f:

$$d_{\mathscr{I},f}([\lambda x \Box(Px \to \sim Qx)]) = \mathscr{NEC}(d_{\mathscr{I},f}([\lambda x\, Px \to \sim Qx]))$$
$$d_{\mathscr{I},f}([\lambda xvwy\, Px \to Qy]) = \mathscr{VAC}_2(d_{\mathscr{I},f}([\lambda xwy\, Px \to Qy]))$$
$$d_{\mathscr{I},f}([\lambda xwy\, Px \to Qy]) = \mathscr{VAC}_2(d_{\mathscr{I},f}([\lambda xy\, Px \to Qy]))$$
$$d_{\mathscr{I},f}([\lambda xy\, Px \to Qy]) = \mathscr{COND}(d_{\mathscr{I},f}(P), d_{\mathscr{I},f}(Q))$$
$$d_{\mathscr{I},f}([\lambda y \Box Gb]) = \mathscr{VAC}_1(\mathscr{NEC}(\mathscr{PLUG}_1(d_{\mathscr{I},f}(G), d_{\mathscr{I},f}(b))))$$
$$d_{\mathscr{I},f}(\Box(E!b \ \&\ Gb)) = \mathscr{NEC}(\mathscr{NEG}(\mathscr{COND}(\mathscr{PLUG}_1(d_{\mathscr{I},f}(E!), d_{\mathscr{I},f}(b)), \mathscr{NEG}(\mathscr{PLUG}_1(d_{\mathscr{I},f}(G), d_{\mathscr{I},f}(b)))))).$$

C. SATISFACTION

If we are given an interpretation $\mathscr{I}$, and an $\mathscr{I}$-assignment f, we may define f **satisfies** ϕ **with respect to** world w as follows:

(1) If ϕ is any primitive zero-place term, f satisfies ϕ with respect to w iff $ext_w(d_{\mathscr{I},f}(\phi)) = T$

(2) If $\phi = \rho^n o_1 \ldots o_n, f$ satisfies ϕ with respect to w iff $(\exists o_1)\ldots(\exists o_n)$ $(\exists r^n)(o_1 = d_{\mathscr{I},f}(o_1)\ \&\ldots\&\ o_n = d_{\mathscr{I},f}(o_n)\ \&\ r^n = d_{\mathscr{I},f}(\rho^n)$ $\&\ \langle o_1, \ldots, o_n\rangle \in ext_w(r^n))$

(3) If $\phi = o\rho^1, f$ satisfies ϕ with respect to w iff

$(\exists o)(\exists r^1)(o = d_{\mathscr{I},f}(o)\ \&\ r^1 = d_{\mathscr{I},f}(\rho^1)\ \&\ o \in ext_{\mathscr{A}}(r^1))$

(4) If $\phi = (\sim\psi), f$ satisfies ϕ with respect to w iff f fails to satisfy ψ with respect to w

(5) If $\phi = (\psi \rightarrow \chi)$, $\mathscr{f}$ satisfies ϕ with respect to $\mathscr{w}$ iff $\mathscr{f}$ fails to satisfy ψ with respect to $\mathscr{w}$ or $\mathscr{f}$ satisfies χ with respect to $\mathscr{w}$

(6) If $\phi = (\forall\alpha)\psi$, $\mathscr{f}$ satisfies ϕ with respect to $\mathscr{w}$ iff $(\forall \mathscr{f}')(\mathscr{f}' \overset{=}{_\alpha} \mathscr{f} \rightarrow \mathscr{f}'$ satisfies ψ with respect to $\mathscr{w})$

(7) If $\phi = (\Box\psi)$, $\mathscr{f}$ satisfies ψ with respect to $\mathscr{w}$ iff $(\mathscr{w}')(\mathscr{f}$ satisfies ψ with respect to $\mathscr{w}')$.

D. TRUTH UNDER AN INTERPRETATION

ϕ is **true under** interpretation $\mathscr{I}$ iff every $\mathscr{I}$-assignment $\mathscr{f}$ satisfies ϕ with respect to $\mathscr{w}_0$. ϕ is **false under** $\mathscr{I}$ iff no $\mathscr{I}$-assignment $\mathscr{f}$ satisfies ϕ with respect to $\mathscr{w}_0$. The definitions of **valid** (i.e., **logically true**) and **model** remain the same.

3. THE LOGIC

A. LOGICAL AXIOMS

The logical axiom schemata fall into five groups: the propositional schemata, the quantificational schemata, the modal schemata, the schemata governing λ-expressions, and the schemata governing descriptions. The presentation of the schemata governing λ-expressions will be interrupted by two definitions, in terms of which the second λ-schema will be constructed. We define a modal closure of ϕ to be any formula obtained by prefixing any finite number of (or possibly zero) boxes to ϕ. Then, with the exception of the schemata governing descriptions, all the modal closures of any instance of the following schemata shall be the logical axioms of our system:

Propositional Schemata

LA1: $\phi \rightarrow (\psi \rightarrow \phi)$

LA2: $(\phi \rightarrow (\psi \rightarrow \chi)) \rightarrow ((\phi \rightarrow \psi) \rightarrow (\phi \rightarrow \chi))$

LA3: $(\sim\phi \rightarrow \sim\psi) \rightarrow ((\sim\phi \rightarrow \psi) \rightarrow \phi)$.

Quantificational Schemata

LA4: (a) $(\alpha)\phi \rightarrow \phi^{\tau}_{\alpha}$, where τ contains no descriptions and is substitutable for α

(b) $(\alpha)\phi \to (\psi^{\tau}_{\beta} \to \phi^{\tau}_{\alpha})$, where ψ is any atomic formula, and τ both contains a description and is substitutable for α, β

LA5: $(\alpha)(\phi \to \psi) \to (\phi \to (\alpha)\psi)$, provided α is not free in ϕ.

Modal Schemata

LA6: $\Box\phi \to \phi$

LA7: $\Box(\phi \to \psi) \to (\Box\phi \to \Box\psi)$

LA8: $\Diamond\phi \to \Box\Diamond\phi$

LA9: $\Box(\alpha)\phi \equiv (\alpha)\Box\phi$

LA10: $(x)(F)(\Diamond xF \to \Box xF)$.

λ-Schemata

λ-EQUIVALENCE: where ϕ is any propositional formula containing no descriptions, the following is an axiom:

$$(x_1)\dots(x_n)([\lambda v_1 \dots v_n\phi]x_1 \dots x_n \equiv \phi^{x_1,\dots,x_n}_{v_1,\dots,v_n}).$$

To more efficiently state the second λ-schema, λ-IDENTITY, we utilize the following two definitions:

D_3 $\quad F^1 = G^1 =_{df} \quad (x)(xF^1 \equiv xG^1)$

D_4 $\quad F^n = G^n =_{df} \quad$ (where $n > 1$)

$$(x_1)\dots(x_{n-1})([\lambda y\, F^n yx_1 \dots x_{n-1}] = [\lambda y\, G^n yx_1 \dots x_{n-1}]\ \&\ [\lambda y\, F^n x_1 yx_2 \dots x_{n-1}] = [\lambda y\, G^n x_1 yx_2 \dots x_{n-1}]\ \&\dots\&\ [\lambda y\, F^n x_1 \dots x_{n-1}y] = [\lambda y\, G^n x_1 \dots x_{n-1}y]).$$

Given D_3 and D_4, we have:

λ-IDENTITY: where ρ^n is any relation term and $v_1, \dots, v_n, v'_1, \dots, v'_n$ are distinct object variables not free in ρ^0, the following is an axiom:

$$[\lambda v_1 \dots v_n \rho^n v_1 \dots v_n] = \rho^n\ \&\ [\lambda v_1 \dots v_n \rho^0] = [\lambda v'_1 \dots v'_n \rho^0].$$

Description schemata

L-DESCRIPTIONS$_1$: where ψ is any atomic formula or conjunction of atomic formulas, the following is an axiom:

$$\psi^{(\imath x)\phi}_{v} \to (\exists y)(\phi^y_x\ \&\ \psi^y_v).$$

L-DESCRIPTIONS$_2$: where ψ is any atomic formula, the following is an axiom:

$$\psi_v^{(\imath x)\phi} \rightarrow (y)(\phi_x^y \rightarrow \psi_v^y).$$

L-DESCRIPTIONS$_3$: where ψ is any atomic formula with v_1 free and χ is any formula with v_2 free, the following is an axiom:

$$\psi_{v_1}^{(\imath x)\phi} \rightarrow \sim((\exists y)(\phi_x^y \mathbin{\&} \chi_{v_2}^y) \mathbin{\&} (\exists y)(\phi_x^y \mathbin{\&} \sim \chi_{v_2}^y)).$$

The propositional and quantificational schemata have been carried over almost intact from the elementary theory. The exception is LA4, which has been modified to accommodate non-denoting descriptions and terms which may contain such descriptions (for discussion on this matter, see the Appendix to Chapter II, part D). These first five axioms have greater significance than their elementary theoretic counterparts, due to the presence of new kinds of formulas and terms. For example, $\Box P \rightarrow (Q \rightarrow \Box P)$ is an instance of LA1; and $(F)Fc \rightarrow [\lambda y \Box Ty]c)$ is an instance of LA4a.

LA6–LA8 are the standard three propositional modal axioms of S_5. LA9 is the second order version of the Barcan formula. Both our object quantifiers and relation quantifiers are unrestricted (they range over everything in their respective domains). Since boxes are interpreted semantically as universal generalizations over the domain of worlds, commuting a box with a universal quantifier is just as valid as commuting two universal quantifiers. Also, diamonds commute with "existential" quantifiers. LA10 is a new logical axiom which governs the modal logic of encoding. It guarantees that objects encode their properties "rigidly" (should they encode any). That is, if they encode a property at some possible world, they encode that property at all possible worlds. To see that LA10 is logically true, note that the encoding extension ($ext_{\mathscr{A}}$) of a property is not relativized to a world (Section 2, A., clause 7). So the conditions for satisfaction for encoding formulas (Section 2, C., clause 3) are totally independent of the worlds. So if an encoding formula is true at some world, it is true at every world.

Then we have our two λ-schemata. λ-EQUIVALENCE, though it has a minor restriction (see the Appendix to Chapter II, part E, for details), has greater significance than its counterpart in the elementary theory, due to the presence of λ-expressions which are vacuous expansions or necessitations. The first conjunct of λ-IDENTITY has the same significance as its counterpart in the elementary theory, however, the second conjunct of

λ-IDENTITY is new. Intuitively, $[\lambda x_1 F^0]$ and $[\lambda x_2 F^0]$ are both names of the same vacuous property. In every interpretation, they will denote semantically identical properties, so the encoding extensions of such properties must be the same. This is what is asserted by the second conjunct of λ-IDENTITY, except that it generalizes to the case where relations are denoted by λ-expressions with more than one vacuously bound variable.

Finally, we have the three description schemata which we discussed in the Appendix to Chapter II, part F. Since our definite descriptions are rigid designators, instances of these schemata aren't necessarily true. They are paradigm examples of logical truths which aren't necessary.

B. RULES OF INFERENCE

Officially, we need only two rules of inference:

(1) Arrow Elimination ("$\rightarrow$E"): from ϕ and $\phi \rightarrow \psi$, we may infer ψ

(2) Universal Introduction ("UI"): from ϕ, we may infer $(\alpha)\phi$

The notion of **proof, logical theorem,** and **provable from** all carry over from the elementary theory. Using these notions, we may state a restricted version of the rule of necessitation which is easily derivable:[6,7]

Box Introduction ("$\Box$I"): If we are given a proof of ϕ from a set of formulas Γ, then if in this proof ϕ does not depend on any unmodalized formula (i.e., formula not beginning with a box), then $\Gamma \vdash \Box\phi$ (dependence is defined in Chapter I, Section 3).

We will appeal to this rule on numerous occasions in Chapter IV. And we will also use the many standard derived rules of the second order modal predicate calculus. Our derived rules of λ-Introduction and λ-Elimination are formulated as in Chapter I, with the restriction that definite descriptions not occur in ϕ.

Note that the RELATIONS theorem schema (I., Section 3) is now derivable without the restriction that $x_1,\ldots,x_n$ be free in ϕ, but must be restricted to ϕ's which contain no descriptions:

RELATIONS: where ϕ is any propositional formula with no free F^n's and no descriptions, the following is a logical theorem:

$$(\exists F^n)\Box(x_1)\ldots(x_n)(F^n x_1 \ldots x_n \equiv \phi).$$

Of course if ϕ contains a definite description, then by LA4b, instances of the above schema follow from the assumption that some atomic formula containing the description is true.

In addition to the examples of this schema offered in Chapter I, we now have further examples:

(a) $(\exists F)\Box(x)(Fx \equiv \Box Gx)$

(b) $(\exists F)\Box(x)(Fx \equiv \Box(E!x \rightarrow Gx))$

(c) $(\exists F)\Box(x)(Fx \equiv Gb)$

(d) $(\exists F)\Box(x)(Fx \equiv \Box Gb)$

(e) $(\exists F)\Box(x)(y)(Fxy \equiv \Box Gb)$.

(a) tells us that for any property G, there is a property of exemplifying G essentially; (b) tells us that for any property G, there is a property of necessarily exemplifying-G-if-existing; (c) and (d) assert, respectively, that there is a property objects exemplify just in case b exemplifies G and just in case necessarily b exemplifies G; (e) asserts that there is a two-place relation objects bear to one another just in case necessarily b exemplifies G.

Note also that while the following is, strictly speaking, not an instance of RELATIONS, it is nevertheless easily derivable:

PROPOSITIONS: Where ϕ is any propositional formula with no free F^0's and no descriptions, the following is a logical theorem:

$$(\exists F^0)\Box(F^0 \equiv \phi).$$

And we may also define the conditions under which propositions are identical:

$\boldsymbol{D}_5$ $\quad F^0 = G^0 =_{df} [\lambda y\ F^0] = [\lambda y\ G^0]$.

That is, propositions F^0 and G^0 are identical iff the property of being such that F^0 is encoded by all and only the objects encoded by the property of being such that G^0. This definition turns out to be extremely useful in Chapter IV, Section 2, where we prove that there is a unique actual world.

RELATIONS, PROPOSITIONS, $\boldsymbol{D}_4$, and $\boldsymbol{D}_5$ comprise a complete modal theory of n-place relations. It is an important feature of this theory that relations with the same exemplification extensions at each possible world may nevertheless be distinct. For example, it is consistent with our theory that the properties of being an equilateral Euclidean triangle and being an equiangular Euclidean triangle are distinct, even though they have the same exemplification extensions at each possible world. And the properties of being-blue-or-not-blue and being-green-or-not-green may be distinct, though logically equivalent.

We call the metaphysical system which consists of the interpreted modal language, together with its logic, the **modal object calculus (with propositions, and complex terms).**

4. THE PROPER AXIOMS

We have again embedded our primitive metaphysical notions in the atomic formulas of the language, and embedded the primitive logical notions (including the new primitive, ***necessarily***) in the complex formulas and terms. We now use our primitive theoretical relations (***existence*** and ***E-identity***), to state the theory of abstract objects (and in the course of doing so, produce a theory of identity as well). The theory has five axioms, three of which are schematic. We assert that the modal closures of the first two unschematic axioms, the modal closures of all the instances of the first two schemata (IDENTITY and *A*-OBJECTS), and the unmodalized instances of the third schema (DESCRIPTIONS) are all true A PRIORI:

AXIOM 1. ("*E*-IDENTITY"): $x=_E y \equiv \Diamond E!x \;\&\; \Diamond E!y \;\&\; \Box(F)(Fx \equiv Fy)$.

AXIOM 2. ("NO-CODER"): $\Diamond E!x \rightarrow \Box \sim(\exists F)xF$.

In order to state the third axiom, we need the following definition:

$$\boldsymbol{D}_6 \qquad x=y=_{df}x=_E y \vee (A!x \;\&\; A!y \;\&\; \Box(F)(xF \equiv yF)).$$

Since the definienda in $\boldsymbol{D}_3$–$\boldsymbol{D}_6$ all have a special logical form, we have:

AXIOM 3. ("IDENTITY"): $\alpha=\beta \rightarrow (\phi(\alpha,\alpha) \equiv \phi(\alpha,\beta))$, where $\phi(\alpha,\beta)$ is the result of replacing some, but not necessarily all, free occurrences of α by β in ϕ (α,α), provided β is substitutable for α in the occurrences of α it replaces.

AXIOM 4. ("*A*-OBJECTS"): for any formula ϕ where x is not free, the following is an axiom:

$$(\exists x)(A!x \;\&\; (F)(xF \equiv \phi)).$$

AXIOM 5. ("DESCRIPTIONS"): where ψ is any atomic or defined object identity formula with one free object variable v, the following is an axiom:

$$\psi_v^{(\imath x)\phi} \equiv (\exists ! y)\phi_x^y \ \& \ (\exists y)(\phi_x^y \ \& \ \psi_v^y).$$

Given our discussion of the axioms and theorems of the elementary theory, these axioms should be straightforward. Semantically, each possible world will look somewhat like a model of elementary object theory. At each world, there are objects which exist there and which fail to exist there. But from the point of view of a given world, say the actual world, the objects which fail to exist divide up into two mutually exclusive classes – the objects which necessarily fail to exist and the objects which exist at some other possible world. So from the point of view of the actual world, E-IDENTITY and NO-CODER govern the objects which either exist at this world or exist at some other world.

The IDENTITY axiom has greater significance than its counterpart in the elementary theory. This is due to the presence of the many new kinds of terms in the language. The following are both instances of IDENTITY:

$$F^0 = G^0 \rightarrow (\Box F^0 \equiv \Box G^0)$$
$$F^1 = G^1 \rightarrow ([\lambda y \Box Fa]b \equiv [\lambda y \Box Ga]b).$$

A-OBJECTS also has greater significance since it now yields objects which encode vacuous and modal properties. Since the modal closures of A-OBJECTS are axioms, the following counts as an axiom schema:

$$\Box(\exists x)(A!x \ \& \ (F)(xF \equiv \phi)), \text{ where } \phi \text{ has no free } x\text{'s}.$$

Semantically, this tells us that given a world ω and a condition on properties ϕ, there is an abstract object at ω which encodes just the properties satisfying ϕ at ω. A formula like "Fs" ("Socrates exemplifies F") is satisfied by different properties at different worlds. At each world, then, there is an A-object which encodes just the properties Socrates exemplifies at that world. A formula like "$F = R \vee F = S$" is satisfied by the same two properties, roundness and squareness, at each world. Given LA10, and definition of identity, the$_{\mathscr{A}}$ round square of one world will be identical with the$_{\mathscr{A}}$ round square of any other world. Intuitively, all of the A-objects from each of the worlds can be grouped into one set, the set of A-objects, which stays fixed from world to world. In the future, when we use restricted z-variables, they will range over this set.

DESCRIPTIONS has been added to our list of axioms and it has a few interesting and important features we should consider. Its instances are paradigm cases of A PRIORI truths which are not necessarily true. Our axiom guarantees that the descriptions in our language that appear in atomic or object identity formulas behave according to our A PRIORI intuitions (see note 1 of Chapter II for reasons why instances of DESCRIPTIONS are not logically true). But it is easy find worlds such that the left side of a given instance of DESCRIPTIONS is true there while its right side is false there. That is because the descriptions of our language are rigid designators. The left side of a given instance of DESCRIPTIONS, which will say essentially "the thing which satisfies ϕ ψ's", will be true at a world w_1 just in case there is a unique object satisfying ϕ at the base world w_0 which satisfies ψ at w_1. But there need not be an object which satisfies both ϕ and ψ at w_1 or which uniquely satisfies ϕ at w_1. But that's what it would take for the right side of the instance of DESCRIPTIONS to be true at w_1. So DESCRIPTIONS is not necessarily true, and given our restricted version of the rule of necessitation, we cannot produce the modalized instances of DESCRIPTIONS as (proper) theorems (for further discussion on this matter, see note 6).

E-IDENTITY, NO-CODER, IDENTITY, *A*-OBJECTS, and DESCRIPTIONS jointly constitute **the modal theory of abstract objects**. Evidence for thinking that the theory is consistent may be found in Appendix A, part C, where the reader will find an extensional model of the monadic portion of the theory described in ZF. It is provable that some propositions as well as some complex relations do not have unique constituents.[8] But such a result might seem insignificant when compared to the potential the theory has for applications.

In these applications, it will be important to distinguish three senses of the phrase "possible object". On one sense of this phrase, objects which satisfy $\boldsymbol{D}_2$ (Section 1) are possible objects, whereas abstract objects are not. We always use "possibly existing object" to indicate this sense of "possible object".

The other two senses of the phrase are ones in which abstract objects are possible objects. Consider $\boldsymbol{D}_7$, where "z" is a restricted variable ranging over abstract objects:

$\boldsymbol{D}_7$ *z is strongly possible* ("$SPoss(z)$") $=_{df} \Diamond(\exists x)(F)(zF \equiv Fx)$.

We always use the phrase "strongly possible object" to indicate this sense of "possible object". For example, Socrates' blueprint is strongly possible, and so is the blueprint of Socrates' blueprint.[9]

The third sense of "possible object" we distinguish requires a preliminary definition.

$\boldsymbol{D}_8$ *x is weakly correlated with z* ("$WCor(x,z)$") $=_{df} (F)(zF \rightarrow Fx)$.

For example, abstract objects which encode just some of the properties a given object exemplifies are incomplete blueprints of the object – the object is weakly correlated with them. We now have,

$\boldsymbol{D}_9$ *z is weakly possible* ("$WPoss(z)$") $=_{df} \Diamond(\exists x)\, WCor(x,z)$.

Weakly possible A-objects are "possible objects" in the sense that they do not encode any contradictory properties. F and G are contradictory properties iff it is not possible that some object exemplify both of them.

The notion of weak correlation we defined in $\boldsymbol{D}_8$ was used in Chapter II, Section 3. Recall that we defined existence$_2$ for an object x ("$E!!x$") as $(\exists y)(E!y \;\&\; (F)(xF \rightarrow Fy))$. So abstract objects can "exist" in the sense that they have a weak correlate which exists. To say that an abstract object x "might have existed" is to say that it is possible that x have a weak correlate which exists ("$\Diamond E!!x$").

We shall keep these distinctions straight in the applications which follow. To prepare for these applications, we add to our primitive vocabulary the usual abbreviations of standard English gerunds. Also, we adopt a modal version of our AUXILIARY HYPOTHESIS – A-objects necessarily fail to exemplify nuclear relations.[10]

CHAPTER IV

THE APPLICATIONS OF THE MODAL THEORY

In much of this chapter, we shall be speaking in the object language. When doing so, everything we say may be analyzed in terms of our four metaphysical primitives (object, n-place relation, exemplifies, encodes), six logical primitives (not, if-then, every, necessarily, the, being such that), and two primitive theoretical relations (exists, E-identical). All of the definitions constructed and theorems proved in what follows may be ultimately analyzed in terms of these primitives. We begin with a definition of truth.

1. TRUTH

Since propositional formulas are also terms which denote propositions, we shall follow Ramsey in supposing that the predicate "is true" and the operator "it is true that" are definable by elimination.[1] The language we developed in the previous chapter allows us to incorporate Ramsey's suggestion through the formulation of the following definitions:

$\boldsymbol{D}_{10}$ $\quad F^0$ *is true* $=_{df} F^0$

$\boldsymbol{D}_{11}$ $\quad$ *It is true that* $F^0 =_{df} F^0$.

Ramsey's idea works fine as long as we are interested in just the truths relative to our world. A less mundane notion of truth is the notion of ***truth at*** a particular world. We shall produce a definition of this notion in the next section, once we have modelled possible worlds. But before we do so, we require a few more preliminary definitions.

We shall say that a property F^1 ***is constructed out of*** a proposition F^0 iff F^1 is the property of being such that F^0:

$\boldsymbol{D}_{12}$ $\quad F^1$ *is constructed out of* F^0 ("$Const(F^1, F^0)$") $=_{df}$ $F^1 = [\lambda x\, F^0]$.

We then define a ***vacuous*** property to be one which is constructed out of some proposition:

$\boldsymbol{D}_{13}$ $\quad F^1$ *is a vacuous property* ("$Vac(F^1)$") $=_{df} (\exists F^0) Const(F^1, F^0)$.

Examples of vacuous properties are: being such that Carter is President, being such that Fischer defeated Spassky, being such that Nixon did not resign the Presidency, being such that a Luxembourgian was the first man on the moon, being such that every man loves every fish, etc. λ-EQUIVALENCE guarantees that necessarily, an object x exemplifies a vacuous property like $[\lambda y F^0]$ iff F^0 is true. So if F^0 is true, everything exemplifies $[\lambda y F^0]$, and if F^0 is not true, nothing does. Consequently, vacuous properties are either "full" (everything exemplifies them) or "empty" (nothing exemplifies them). Indeed, some properties will be necessarily full and others will be necessarily empty. Being such that either Carter is President or Carter is not President ($[\lambda y\, Pc \vee \sim Pc]$) is an example of the former; being such that both Carter is President and Carter is not President ($[\lambda y\, Pc \,\&\, \sim Pc]$) is an example of the latter.

Finally, it will be important to define conditions under which we can say that an abstract object encodes a proposition. Consider $\boldsymbol{D}_{14}$:

$\boldsymbol{D}_{14}$ *z encodes* F^0 ("$\Sigma_z F^0$") $=_{df} z[\lambda y F^0]$.

That is, an abstract object z encodes a proposition F^0 iff z encodes being such that F^0.

2. MODELLING POSSIBLE WORLDS[2]

Possible worlds will be abstract objects which encode only vacuous properties and which meet two other conditions. For one thing, they must be ***maximal***, i.e., for every proposition F^0, either they encode F^0 or they encode the negation of F^0.

$\boldsymbol{D}_{15}$ *z is maximal* ("$Max(z)$") $=_{df} (F^0)(\Sigma_z F^0 \vee \Sigma_z \sim F^0)$.

So if an object z is maximal, it must encode, for every proposition F^0, either being such that F^0 or being such that it is not the case that F^0.

Secondly, worlds must in some sense be possible objects. One way to make this requirement precise would be to stipulate that worlds must be weakly possible (i.e., as in III, Section 4, $\boldsymbol{D}_9$). This would require that it be possible that some object exemplifies every property the world encodes. However, a more elegant way of ensuring that inconsistent propositions will not be encoded by the same world is to stipulate that if an object z is to be a world, then it must be possible that every proposition z encodes is true.[3]

We can formalize all these conditions on worlds in the following definition:

z *is a possible world* $=_{df}(F^1)(zF \rightarrow Vac(F))$ & $Max(z)$ & $\Diamond(F^0)(\Sigma_z F^0 \rightarrow F^0)$.

Although this definition would serve us well, there is a more elegant definition which is equivalent:

$\boldsymbol{D}_{16}$ z *is a possible world* ("*World*(z)") $=_{df}(F)(zF \rightarrow Vac(F))$ & $\Diamond(F^0)(\Sigma_z F^0 \equiv F^0)$.

That is, an object z is a world iff every property it encodes is vacuous and it is possible that z encodes all and only true propositions. Given $\boldsymbol{D}_{16}$, we can prove that worlds are maximal:

THEOREM 1. $(z)(World(z) \rightarrow Max(z))$.

Proof. Suppose z_5 is an arbitrary world. By definition, $\Diamond(F^0)(\Sigma_{z_5} F^0 \equiv F^0)$. We want to conclude that for an arbitrary proposition Q^0, that $\Sigma_{z_5} Q^0$ or $\Sigma_{z_5} \sim Q^0$. We do this in two stages: in stage (A), we prove that $\Diamond(\Sigma_{z_5} Q^0 \vee \Sigma_{z_5} \sim Q^0)$, and in stage (B), we use a theorem of S_5 (which distributes a $\Diamond$ over a disjunction) and our new logical axiom LA10, to prove that $\Sigma_{z_5} Q^0$ or $\Sigma_{z_5} \sim Q^0$.

(A) In this stage, we rely on the following theorem of S_5: $\Box(\phi \rightarrow \psi) \rightarrow (\Diamond\phi \rightarrow \Diamond\psi)$. If we let $\phi = \ulcorner(F^0)(\Sigma_{z_5} F^0 \equiv F^0)\urcorner$ and $\psi = \ulcorner\Sigma_{z_5} Q^0 \vee \Sigma_{z_5} \sim Q^0\urcorner$, then by establishing that $\Box(\phi \rightarrow \psi)$, we can apply the S_5 theorem using the fact that $\Diamond(F^0)(\Sigma_{z_5} F^0 \equiv F^0)$ and reap our initial result. So we first establish that $(\phi \rightarrow \psi)$, and then use $\Box$I. So assume ϕ, and instantiate the quantifier to both Q^0 and $\sim Q^0$. So $\Sigma_{z_5} Q^0 \equiv Q^0$ and $\Sigma_{z_5} \sim Q^0 \equiv \sim Q^0$. Since $Q^0 \vee \sim Q^0$, it follows that $\Sigma_{z_5} Q^0 \vee \Sigma_{z_5} \sim Q^0$, i.e., ψ. So $\Box(\phi \rightarrow \psi)$, by $\Box$I. And by the S_5 theorem, we have our initial result: $\Diamond(\Sigma_{z_5} Q^0 \vee \Sigma_{z_5} \sim Q^0)$.

(B) It is also a theorem of S_5 that $\Diamond(\phi \vee \psi) \rightarrow (\Diamond\phi \vee \Diamond\psi)$. By letting ϕ, ψ be the disjuncts of our initial result, it follows that $\Diamond\Sigma_{z_5} Q \vee \Diamond\Sigma_{z_5} \sim Q^0$. By LA10, it follows that $\Box\Sigma_{z_5} Q^0 \vee \Box\Sigma_{z_5} \sim Q^0$, since if possibly an object encodes a property, it does so necessarily. And by another theorem of S_5, $\Box(\Sigma_{z_5} Q^0 \vee \Sigma_{z_5} \sim Q^0)$. By LA6, $\Sigma_{z_5} Q^0 \vee \Sigma_{z_5} \sim Q^0$, i.e. $Max(z_5)$. So every $\boldsymbol{D}_{16}$-world is maximal. ⊠

The proof of Theorem 1 could be simplified a great deal if we think model-theoretically, using the notion of a possible world as a primitive. If we signify that we are appealing to the semantically primitive notion of a world by shifting type styles and writing "*world*", the proof of Theorem 1 would read as follows:

Suppose z_5 is an arbitrary world. By $\boldsymbol{D}_{16}$, at some possible *world*, say w_1, z_5 encodes there all and only the propositions true there. For any proposition F^0, either F^0 is true at w_1 or $\sim F^0$ is true at w_1. So for all propositions F^0, either z_5 encodes F^0 at w_1 or z_5 encodes $\sim F^0$ at w_1. But by LA10, if an abstract object encodes a property at some possible *world*, it encodes it at all possible *worlds*. So (at the *actual world*, w_0) for all propositions F^0, either z_5 encodes F^0 or z_5 encodes $\sim F^0$. So every world is maximal. ⊠

Readers who find the semantically primitive notion of a *world* an intuitive one will find it much easier to construct model theoretic proofs like the above for the theorems which follow. The proofs are a good deal simpler, since the theorems turn out to be almost immediate consequences of the axioms and definitions. However, in the context of the present work, the presentation of such model-theoretic proofs has a curious disadvantage. It fosters the wrong impression. It would encourage the reader to suppose that there are *A*-objects which can represent the worlds. But the POINT of the theorems which follow is to show that the worlds just ARE *A*-objects. Consequently, I shall not present any further proofs model-theoretically.[4] Those who prefer to think model-theoretically need not attend to the proofs offered in the text to verify that the following claims are in fact theorems. But these readers should not suppose that the model-theoretic results could be considered to be a substitute for the metaphysical results. They are simply a device for quickly verifying that the metaphysical claims do follow.

Theorem 1 is instrumental for showing that $\boldsymbol{D}_{16}$ implies our first definition of 'possible world'. It is a straightforward proof-theoretic exercise to show that the first definition implies $\boldsymbol{D}_{16}$.[5]

Let us say that propositions F^0 and G^0 are ***inconsistent*** iff it is not possible that both F^0 and G^0 be true. We may then say that an *A*-object z is consistent iff it is not the case that z encodes inconsistent propositions:

$\boldsymbol{D}_{17}$ *z is consistent* ("*Cons* (*z*)") $=_{df}$ $\sim(\exists F^0)(\exists G^0)(\sim\Diamond(F^0 \;\&\; G^0) \;\&\; \Sigma_z F^0 \;\&\; \Sigma_z G^0)$.

We then have:

THEOREM 2. $(z)(World(z) \rightarrow Cons(z))$.
That is, every world is consistent.

Proof. Assume for reductio that $World(z_4)$, $\sim\Diamond(P^0 \;\&\; Q^0)$, $\Sigma_{z_4}P^0$, and

$\Sigma_{z_4}Q^0$. Note that by $\Diamond$I and LA10, these last two assumptions must be necessary truths, i.e. $\Box\Sigma_{z_4}P^0$ and $\Box\Sigma_{z_4}Q^0$. So by a theorem of S_5, $\Box(\Sigma_{z_4}P^0 \;\&\; \Sigma_{z_4}Q^0)$. Now by $\boldsymbol{D}_{16}$, $\Diamond(F^0)(\Sigma_{z_4}F^0 \equiv F^0)$. Now let $\phi = \ulcorner(F^0)(\Sigma_{z_4}F^0 \equiv F^0)\urcorner$ and $\psi = \ulcorner P^0 \;\&\; Q^0\urcorner$. We shall want to show that $\Box(\phi \rightarrow \psi)$. Clearly, $\Box((\Sigma_{z_4}P^0 \;\&\; \Sigma_{z_4}Q^0) \rightarrow (\phi \rightarrow \psi))$. But since $\Box(\Sigma_{z_4}P^0 \;\&\; \Sigma_{z_4}Q^0)$, it follows by a theorem of S_5 (namely, that $\Box(\chi \rightarrow \theta) \;\&\; \Box\chi \rightarrow \Box\theta)$ that $\Box(\phi \rightarrow \psi)$. Now by the using S_5 theorem that $\Box(\phi \rightarrow \psi) \rightarrow (\Diamond\phi \rightarrow \Diamond\psi)$, it follows that $\Diamond(P^0 \;\&\; Q^0)$, contrary to hypothesis. ⊠

Another result quickly falls out of our definitions:

THEOREM 3. $(z)(World(z) \rightarrow (F^0)(G^0)((\Sigma_z F^0 \;\&\; F^0 \Rightarrow G^0) \rightarrow \Sigma_z G^0))$. That is, all the necessary consequences of propositions encoded in a world are also encoded in that world.

Proof. Assume $World(z_3)$, $\Sigma_{z_3}P^0$, and that $P^0 \Rightarrow Q^0$ (i.e., $\Box(P^0 \rightarrow Q^0)$). Again, $\Box\Sigma_{z_3}P^0$, by $\Diamond$I and LA10. So if $\phi = \ulcorner(F^0)(\Sigma_{z_3}F^0 \equiv F^0)\urcorner$ and $\psi = \ulcorner\Sigma_{z_3}Q^0\urcorner$, we easily get $\Box(\phi \rightarrow \psi)$ by theorems of S_5. So by again appealing to the theorem that $\Box(\phi \rightarrow \psi) \rightarrow (\Diamond\phi \rightarrow \Diamond\psi)$, we get $\Diamond\Sigma_{z_3}Q^0$, since $\Diamond\phi$ follows from the definition of a world. So by LA10 and LA6, $\Sigma_{z_3}Q^0$. ⊠

Theorems 1, 2, and 3 should give us a good grasp on the inner workings of the theory, as well as the import of the second clause of $\boldsymbol{D}_{16}$. They should also help us to see that the following definition is justified:

$\boldsymbol{D}_{18}$ $\quad F^0$ *is true at* $z =_{df} World(z) \;\&\; \Sigma_z F^0$.

So whenever z is a world, the propositions true at z are just the propositions z encodes. This definition suggests what it is for a world to be actual:

$\boldsymbol{D}_{19}$ $\quad z$ *is an actual world* ("$World_\alpha(z)$") $=_{df} World(z) \;\&\; (F^0)(\Sigma_z F^0 \equiv F^0)$.

That is, an actual world is any world such that all and only true propositions are true at that world. We now get the following result:

THEOREM 4. $(z)(z')(World_\alpha(z) \;\&\; World_\alpha(z') \rightarrow z' = z)$.
That is, there is at most one actual world.

Proof. Suppose, for reductio, that $World_\alpha(z_1)$ and $World_\alpha(z_2)$, where $z_1 \neq z_2$. Since z_1 and z_2 are distinct *A*-objects, they must differ with respect to at least one encoded property. Since they are both worlds, any such property must be vacuous. So without loss of generality, suppose $\Sigma_{z_1}Q^0$ &

$\sim\Sigma_{z_2}Q^0$. By Theorem 1, z_2 must be maximal. So $\Sigma_{z_2}\sim Q^0$. But since both z_1 and z_2 are actual, every proposition they encode must be true. Contradiction. ☒

We also get:

THEOREM 5. $(\exists z)World_\alpha(z)$.

Proof. By *A*-OBJECTS, there is an abstract object which encodes a property F iff it's a vacuous property constructed out of a true proposition, i.e., $(\exists z)(F)(zF \equiv (\exists F^0)(F^0 \,\&\, F = [\lambda y F^0]))$. Call this object z_0. To show z_0 is an actual world, we show that it satisfies both clauses of $\boldsymbol{D}_{19}$. So we show (a) $World(z_0)$, and (b) $(F^0)(\Sigma_{z_0}F^0 \equiv F^0)$:

(a) Clearly, every property z_0 encodes will be vacuous. So we want to show that possibly, all and only the true propositions are encoded in z_0. Consider an arbitrary proposition Q^0. $(\rightarrow)$ Suppose $\Sigma_{z_0}Q^0$. Then by definition of z_0, $(\exists F^0)(F^0 \,\&\, [\lambda y Q^0] = [\lambda y F^0])$. Call this proposition R^0. Since $[\lambda y Q^0] = [\lambda y R^0]$, it follows from the definition of proposition identity (III, Section 3, $\boldsymbol{D}_5$) that $Q^0 = R^0$. Since R^0 is true, so is Q^0. $(\leftarrow)$ Suppose Q^0 is true. Then $z_0[\lambda y\ Q^0]$, i.e., $\Sigma_{z_0}Q^0$.

Since we have established that $\Sigma_{z_0}Q^0 \equiv Q^0$, for an arbitrary proposition Q^0, it follows that $\Diamond(F^0)(\Sigma_{z_0}F^0 \equiv F^0)$.

(b) Clearly, z_0 encodes all and only the propositions which are true, as we have just shown. ☒

With Theorems 4 and 5, we have proven that there is a unique actual world (from A PRIORI assumptions alone). We are entitled to name this object and we do so as follows: $w_\alpha =_{df} (\iota z)World_\alpha(z)$.

It should be interesting that there is an actual world even though it does not exist. If we had proven that the actual world existed, then we would have just proven that something exists from A PRIORI assumptions; yet A PRIORI theories should not have contingent consequences. Also, if the actual world were an existing object like you, me, or some sub-atomic particle, it would fail to encode any properties (by NO-CODER). There would be no reason to think that its vacuous properties were any more crucial to its identity than other properties it exemplified (like not being a cat, being non-red). Recall here Wittgenstein's dictum that the world is just all that is the case.[6] These considerations make it easy to see that the word "actual" as it occurs in the English phrase "the actual world" does not mean "existing". (It is intriguing that the word "is" in

Wittgenstein's dictum may again be read as "encodes", since given our derivative use of this notion, w_α encodes just all that is the case.)

These remarks on existence and actuality bring us to the following, important definition:

$\boldsymbol{D}_{20}$ $\quad$ *x exists at z* $=_{df}$ *World* (z) & $\Sigma_z E!x$.

That is, an object x exists at a world iff the proposition that x exists is true at that world. Note that we cannot prove from our A PRIORI assumptions that anything exists at w_α. This is a fortunate result since, as we noted earlier, it must surely be a contingent matter that something exists (at the actual world). Let us say that a world of existing things is any actual world such that something exists:

$\boldsymbol{D}_{21}$ $\quad$ *z is a world of existing things* $=_{df}$ $World_\alpha(z)$ & $(\exists x)E!x$.

Clearly, there could be at most one world of existing things – all such worlds would have to be identical with w_α. If we add the contingent assumption that something exists, it follows that w_α is THE world of existing things. So we must add a contingent assumption to prove that there is a unique world of existing things, though it is true A PRIORI that there is a unique actual world.

We might note in passing that it is consistent with our definition of existing at a world that objects exist at more than one world. However, some philosophers apparently like to work with a notion of existing at a world on which objects can exist at at most one world.[7] We could accommodate the views of these philosophers were we to define existence at a world as follows:

$$x \text{ exists at } z =_{df} World(z) \ \& \ (F^0)(\Sigma_z F^0 \equiv [\lambda y F^0]x).$$

Using this definition, we would get the result that individuals are world-bound. For suppose b exists at worlds z_1 and $z_2, z_1 \neq z_2$. If $[\lambda y\, Q]$ was the vacuous property distinguishing z_1 and z_2, it would follow that b both exemplified and failed to exemplify this property. So b cannot exist at both, on this definition of "exists at". Counterpart theorists may then prefer to use this latter definition in their investigations.

One of the most important theorems to fall out of our theory verifies a now common philosophical intuition – necessary propositions are true at all possible worlds. Let "w" be a restricted variable ranging over the worlds:

THEOREM 6. $(F^0)(\Box F^0 \equiv (w)\Sigma_w F^0)$.

Proof. ($\rightarrow$) Assume $\Box Q^0$, where Q^0 is arbitrary and show $\Sigma_{w_1} Q^0$, where w_1 is arbitrary. Let $\phi = \ulcorner (F^0)(\Sigma_{w_1} F^0 \equiv F^0) \urcorner$ and $\psi = \ulcorner \Sigma_{w_1} Q^0 \urcorner$, and by now familiar reasoning, it follows that $\Diamond \Sigma_{w_1} Q^0$, using the fact that $\Box(\phi \rightarrow \psi) \rightarrow (\Diamond \phi \rightarrow \Diamond \psi)$. So $\Sigma_{w_1} Q^0$, by LA10 and LA6. ($\leftarrow$) Assume $(w)\Sigma_w Q^0$, for an arbitrary Q^0, i.e., $(x)(World(x) \rightarrow \Sigma_x Q^0)$. Let us show that this must then be necessary. By the Barcan formulas, it suffices to show: $(x)\Box(World(x) \rightarrow \Sigma_x Q^0)$. Suppose not. Then $\Diamond(World(b) \;\&\; {\sim}\Sigma_b Q^0)$, where "$b$" is arbitrary. Hence, $\Diamond World(b) \;\&\; \Diamond{\sim}\Sigma_b Q^0$. Since $\Diamond World(b) \rightarrow World(b)$ (exercise) and $\Diamond{\sim}\Sigma_b Q^0 \rightarrow {\sim}\Sigma_b Q^0$ (LA10), we get $World(b) \;\&\; {\sim}\Sigma_b Q^0$. So, ${\sim}(World(b) \rightarrow \Sigma_b Q^0)$, contrary to hypothesis. Hence, $\Box(w)\Sigma_w Q^0$. Now if we can show that $\Box((w)\Sigma_w Q^0 \rightarrow Q^0)$, then by LA7, we are done. Well if $(w)\Sigma_w Q^0$, then $\Sigma_{w_\alpha} Q^0$, and by definition of w_α, Q^0. So $(w)\Sigma_w Q^0 \rightarrow Q^0$, and hence $\Box((w)\Sigma_w Q^0 \rightarrow Q^0)$. ⊠

We conclude this section on worlds with a proof of a lemma which will be instrumental in Section 3. Again, let "w" be a restricted variable ranging over the worlds.

LEMMA. $(F^0)(w)(x)(\Sigma_w[\lambda y F^0]x \equiv \Sigma_w F^0)$.
That is, for any object x, x exemplifies being such that F^0 at w iff F^0 is true at w.

Proof. Let Q^0, w_1, and b_1 be an arbitrary proposition, world, and object, respectively. Note that by λ-EQUIVALENCE, $(x)([\lambda y Q^0]x \equiv Q^0)$, and so $[\lambda y Q^0]b_1 \equiv Q^0$. Thus, by $\Box$I, $\Box([\lambda y Q^0]b_1 \equiv Q^0)$, i.e., $[\lambda y Q^0]b_1 \Leftrightarrow Q^0$. ($\rightarrow$) Assume $\Sigma_{w_1}[\lambda y Q^0]b_1$. Then by Theorem 3, it follows that $\Sigma_{w_1} Q^0$. ($\leftarrow$) Assume $\Sigma_{w_1} Q^0$. Then by Theorem 3 again, it follows that $\Sigma_{w_1}[\lambda y Q^0]b_1$. ⊠

3. MODELLING LEIBNIZ'S MONADS[8]

The investigation of monads is as philosophically satisfying as the definition of truth and the investigation of worlds. Although it is unclear what Leibniz intended his monads to be, they have traditionally been regarded as properties of some sort.[9] However, we model them here as abstract objects which are strongly possible (III, Section 4, $\boldsymbol{D}_7$).[10] Strongly possible abstract objects have correlates "at" some possible world. For example, Socrates' blueprint is a monad since it has a correlate at the actual world. Intuitively, "compossible" monads have correlates at the same world. So your blueprint and my blueprint are compossible. A monad "mirrors" the world at which it has a correlate by encoding the vacuous properties the correlate exemplifies – properties constructed out of the propositions true at that world.

To make these ideas precise, we utilize the following definitions. As with the previous lemma, we use "w" as a restricted variable ranging over the abstract objects which satisfy the definition of a world:

D_{22} x *is a correlate of* z *at* w ("$Cor(x, z, w)$") $=_{df} (F)(\Sigma_w Fx \equiv zF)$.

That is, x is a correlate of z at w iff x exemplifies at w exactly the properties z encodes.

D_{23} z *appears at* w ("$Appear(z, w)$") $=_{df} (\exists x) Cor(x, z, w)$.

D_{24} z *is a monad* ("$Monad(z)$") $=_{df} (\exists w) Appear(z, w)$.

D_{25} z *mirrors* w ("$Mirror(z, w)$") $=_{df} (F^0)(\Sigma_w F^0 \equiv \Sigma_z F^0)$.

Using the lemma at the end of Section 2, we now get the following result:

THEOREM 7. $(z)(w)\,(Monad(z)\ \&\ Appear(z, w) \rightarrow Mirror(z, w))$.
That is, every monad mirrors any world where it appears.

Proof. Suppose z_7 is a monad and z_7 appears at w_7. We want to show for an arbitrary proposition Q^0, that $\Sigma_{w_7} Q^0 \equiv \Sigma_{z_7} Q^0$. ($\rightarrow$) Suppose $\Sigma_{w_7} Q^0$. Since z_7 appears at w_7, it has a correlate there. Suppose b_7 is a correlate of z_7 at w_7. So z_7 encodes exactly the properties b_7 exemplifies at w_7. In particular, z_7 encodes $[\lambda y\, Q^0]$ iff b_7 exemplifies $[\lambda y\, Q^0]$ at w_7. By our assumption, $\Sigma_{w_7} Q^0$. So $\Sigma_{w_7} [\lambda y\, Q^0] b_7$, by the lemma proved at the end of Section 2. So $\Sigma_{z_7} Q^0$.

($\leftarrow$) Suppose $\Sigma_{z_7} Q^0$. Again, let b_7 be an object which exemplifies at w_7 exactly the properties z_7 encodes. Clearly b_7 must exemplify $[\lambda y\, Q^0]$ at w_7. So by the lemma, $\Sigma_{w_7} Q^0$. ⊠

Another interesting fact about monads is provable with the help of Theorem 7:

THEOREM 8. $(z)\,(Monad(z) \rightarrow (\exists! w)\, Appear(z, w))$.
That is, every monad appears at a unique world.[11]

Proof. Suppose z_8 is a monad. So there is a world, say w_1, where it appears. We want to show that w_1 is unique, so for reductio, suppose z_8 appears also at w_2, $w_2 \neq w_1$. Since the worlds are distinct, there must be some vacuous property which distinguishes them. Without loss of generality, suppose $\Sigma_{w_1} Q^0$ and $\sim \Sigma_{w_2} Q^0$. And since w_2 is maximal, $\Sigma_{w_2} \sim Q^0$. But by

Theorem 7, z_8 mirrors both worlds. So $\Sigma_{z_8}Q^0$ and $\Sigma_{z_8} \sim Q^0$. But this is impossible, since Q^0 and $\sim Q^0$ would both be true in any world where z_8 has a correlate. ☒

Since we know that every monad appears at a unique world, we are entitled to talk about THE world where it appears. Let us use "*m*" as a restricted variable ranging over the objects satisfying the definition of monad. We then define:

D_{26} $w_m =_{df} (\iota w)\text{Appear}(m, w)$.

Theorems 7 and 8 allow us to say that every monad mirrors its world.[12] Here now is a definition of compossibility.

D_{27} m_1 *is compossible with* m_2 ("$Comp(m_1, m_2)$") $=_{df}$ $(\exists w)(\text{Appear}(m_1, w) \,\&\, \text{Appear}(m_2, w))$.

With these definitions, we have the following lemma:[13]

LEMMA. $(m_1)(m_2)(Comp(m_1, m_2) \equiv w_{m_1} = w_{m_2})$.
That is, two monads are compossible iff the worlds where they appear are identical.

Proof. ($\rightarrow$) Since m_1 and m_2 are compossible, call the world where they both appear w_0. By Theorem 8, $w_0 = w_{m_1}$ and $w_0 = w_{m_2}$. So $w_{m_1} = w_{m_2}$. ($\leftarrow$) Clearly, if the worlds where they appear are identical, there is a world where they both appear. ☒

With the help of this lemma, we get the following result:[14]

THEOREM 9. $(m_1)(m_2)(m_3)(Comp(m_1, m_1) \,\&\, (Comp(m_1, m_2) \rightarrow Comp(m_2, m_1)) \,\&\, (Comp(m_1, m_2) \,\&\, Comp(m_2, m_3) \rightarrow Comp(m_1, m_3)))$.
That is, compossibility is an equivalence notion among the monads.

Proof. Clearly, compossibility is reflexive and symmetrical. To show transitivity, suppose $Comp(m_1, m_2)$ and $Comp(m_2, m_3)$. By the previous lemma, $w_{m_1} = w_{m_2}$ and $w_{m_2} = w_{m_3}$. So $w_{m_1} = w_{m_3}$. ☒

It should also be clear that by defining "embedding" as follows:

D_{28} z_1 *is embedded* in z_2 ("$Embed(z_1, z_2)$") $=_{df} (F^1)(z_1F \rightarrow z_2F)$,

we can prove that every monad has the world where it appears embedded in it:

THEOREM 10. (m) *Embed* (w_m, m).

Proof. m mirrors its world w_m by encoding all the vacuous properties w_m encodes. So w_m must be embedded in m since the vacuous properties exhaust the properties w_m encodes. ☒

Consequently, every monad will be maximal with respect to the propositions. But an even stronger claim is warranted – monads are complete:

$\boldsymbol{D}_{29}$ *z is complete* ("*Com*(z)") $=_{df} (F)(zF \vee z\bar{F})$,

where $\bar{F} =_{df} [\lambda x \sim Fx]$.

THEOREM 11. $(z)(Monad(z) \rightarrow Com(z))$.

Proof. Clearly, if m_0 is a monad, then some object is its correlate at the world where it appears. That object must exemplify there, for every property, either it or its negation. Consequently, m_0 will encode, for every property, either it or its negation. ☒

Theorems (1)–(11) outline a certain picture of objects, monads, and worlds. I believe that this picture is informative, and even insightful, in its own right, independently of any potential it might have for understanding Leibniz. The model has been described at a level of generality which allows us to add a few constraints and investigate the submodels which result. For example, we might want to investigate the structure which results upon adding the hypothesis that there is at most one world where nothings exists. Or we might want to look at the model which results upon adding the hypothesis that abstract objects necessarily exemplify any property they possibly exemplify which is not a relational property with respect to possibly existing objects (the second hypothesis might imply the first, but I do not think they are equivalent). And I think there would be some interest in an investigation of ***E-monads***, i.e., monads which have correlates that exist at the worlds where the monads appear.

Finally, we consider how useful our model is for understanding the work of Leibniz. Leibniz thought propositions were composed of concepts and logical relationships. By concepts (or notions), Leibniz meant properties, things which may or may not be exemplified by individuals. In addition to the general concepts of being human, being red, etc., Leibniz supposed there to be individual concepts. The concept Socrates, the concept Alexander, the concept Adam, etc., are all examples of individual concepts. These are not to be identified with properties like being identical$_E$ to Socrates ($[\lambda x\, x =_E s]$), since these properties have individuals as constituents. Leibniz preferred to develop a calculus of propositions and concepts which did not have individuals as constituents (§12, "Elements of a Calculus").

Leibniz took the logical relationship of concept containment to be crucially involved in the analysis of categorical propositions. Containment is involved in universal affirmative categorical propositions like "every pious man is happy" (§7, "Elements of a Calculus"). The analysis of this sentence is: the concept pious man contains the concept of being happy. Leibniz seems to extend this kind of analysis to singular affirmative propositions. In the "General Inquiries" (§16), he practically identifies the sentential form "A is B" with the form "A contains B". In the "Discourse on Metaphysics" (§8), he analyzes (1),

(1) Alexander is a king,

as: the concept Alexander contains the concept of being a king. He says in this section,

"It is the nature of an individual substance, or complete being to have a notion so complete that it is sufficient to contain, and render deducible from itself, all the predicates of the subject to which this notion is attributed".

And later in the same work (§13) he says,

"the notion of an individual substance contains, once and for all, everything that can happen to it".

Leibniz is not just claiming that the copula "is" should be read as "contains", but rather that the very structure of singular affirmative propositions is a relationship whereby the subject concept contains the predicate concept. And so we find in the "Correspondence with Arnauld" (May 1686), the following data destined for this analysis:

(2) Adam is the first man.
(3) Adam lived in a pleasure garden.
(4) Adam contributed a rib to Eve.
(5) Adam had two sons.

This analysis becomes puzzling when we consider how strong a relation Leibniz intended containment to be. In the "Elements of a Calculus" (§§7, 17), he says that "the subject concept, taken absolutely,..., always contains the predicate concept". Remarks such as this, and the general tone of the discussion in the "Correspondence with Arnauld" leads us to believe that if concept containment holds between two concepts, then it holds in all possible worlds and is not relativized to any circumstance. This puzzled Arnauld, as well as many subsequent philosophers, since it seems to analyze our contingent data (1)–(5) as necessary truths.

It should be clear how we should proceed to unravel the puzzle. The idea is that the Leibnizian concept is a conflation between property and A-object, that proper names are ambiguously used sometimes to speak about the individual and sometimes to speak about the blueprint of the individual, and that encoding should represent containment. We may suppose: (1) that Leibniz's general concepts just are properties, (2) that Leibniz's individual concepts are our monads and in particular, that the concept Adam is just Adam's blueprint, and (3) that Leibniz's analyses utilizing concept containment, in the case of singular affirmative propositions, can be understood in terms of encoding.[15] We make these suppositions precise by constructing the following definitions, utilizing "κ" to range over proper names.

$\boldsymbol{D}_{30}$ *the concept* $F =_{df} F$

$\boldsymbol{D}_{31}$ *z is an individual concept* ("IC(z)") $=_{df}$ *Monad* (z)

$\boldsymbol{D}_{32}$ the concept κ ("$\bar{\kappa}$") $=_{df}$
$(\iota z)(F)(zF \equiv F\kappa)$, i.e.,
(ιz)*Blue* (z,κ)

$\boldsymbol{D}_{33}$ *the concept* κ *contains the concept* $F =_{df} \bar{\kappa}F$.

With these definitions, our theory begins to generate many predictions. For example, it predicts:

THEOREM 12. IC($\bar{\kappa}$).
That is, the concept κ is an individual concept. So we are not just assuming pretheoretically that the concept Adam is an individual concept, but rather we prove it from more general assumptions.

Proof. For an arbitrary property P, we know $\bar{\kappa}P \equiv P\kappa$, by $\boldsymbol{D}_{32}$ and A-DESCRIPTIONS. By the definition of w_α, $P\kappa \equiv \Sigma_{w_\alpha} P\kappa$. So $\bar{\kappa}P \equiv \Sigma_{w_\alpha} P\kappa$. So there is a world where $\bar{\kappa}$ appears, and hence, *Monad* ($\bar{\kappa}$). ☒

The theory also predicts that Leibniz's analyses of (1)–(5) are necessary truths. For example, his analysis of (1) is (1)′:

(1)′ The concept Alexander contains the concept of being a king
By $\boldsymbol{D}_{33}$, this just means:

(1)″ $\bar{a}K$ ("the blueprint of Alexander encodes being a king") By LA10, this is a necessary truth.

The theory also predicts that all individual concepts are complete.

In addition to this positive evidence, there are passages in the letters to Arnauld which support our suppositions. The word "Adam" is sometimes used to talk about Adam ("the actual Adam") and sometimes used to talk about the complete concept of Adam (one of "the many possible Adams"). Leibniz was even aware of this problem (§14, letter, May 1686; §12, letter, July 14, 1686). This ambiguity might also explain why Leibniz calls monads both living and perpetual in §56 of the *Monadology*. This could be symptomatic of conflating the blueprints of persons with the persons themselves. On our understanding of monads, we can see how they could be perpetual. By the AUXILIARY HYPOTHESES, we suppose that monads do not have spatio-temporal location and therefore are not subject to the laws of generation and decay. But it is difficult to see how monads could be thought of as "living".

It also seems appropriate to suggest that our notion of mirroring could serve to represent both Leibniz's notion of mirroring (in the *Monadology*) and his notion of expression (in his letters to Arnauld). Leibniz says repeatedly in the letters to Arnauld that every individual substance expresses (in its concept) the universe into which it enters (May 1686; July 14, 1686). The concepts of individual substances must surely be individual concepts (i.e., monads), and these "express" (i.e., encode all the vacuous properties encoded by) the world into which it enters (i.e., where it appears).

These suggestions should anchor our model in the traditions of Leibnizian scholarship. Any decision about its merit must be the outcome of future discussion. But one word of warning is in order. When we claim that the above model may be useful for "understanding Leibniz' ideas", we are not claiming that the model is what Leibniz intended or had in mind. The point of the exercise, as we see it, is to first theorize about the way the world is and to then apply the theory by predicting some of the things that Leibniz seemed to want to say. The model is to be judged by how well it succeeds in helping us to explain why Leibniz may have said some of the things he in fact said. And in the course of modelling his ideas, we have found further evidence for supposing that the "is" of natural language has a reading on which it means "encodes". Indeed, our work suggests not only that there is a lexical ambiguity in the copula, but also that there may even be a structural ambiguity in the form of singular affirmative prediction itself. In the next section, we find an entire range of data which could be explained by this latter hypothesis.

4. MODELLING STORIES AND NATIVE CHARACTERS

By adding a few primitives to the language of Chapter III, we may model stories, and certain characters in them, as *A*-objects. First we add abbreviations for any proper name of English which denotes an object which, pretheoretically, we judge to be a story (for example, novels, myths, legends, plays, dreams, etc.), or an author or character thereof (where we take characters to be any story object, not necessarily animate). So we shall have object names in our language which abbreviate "*The Tempest*", "Shakespeare", "Prospero", "*The Brothers Karamazov*", "Alyosha", "*The Clouds*", "Strepsiades", "Socrates", "*Ulysses*", "Joyce", "Bloom", "Dublin", etc.

Secondly, we add the name of a new primitive relation which is of central importance to our investigations – the ***authorship*** relation. The formula "Axy" shall say that x authors y, and we trust that our readers have at least an intuitive grasp on what it is to author something.

Consequently, we may define:

$\boldsymbol{D}_{34}$ *z is a story* ("$Story(z)$") $=_{df}$ $(F)(zF \rightarrow Vac(F))$ & $(\exists x)(E!x$ & $Axz)$.

That is, stories are abstract objects which encode only vacuous properties and which are authored by some existing thing. Hence, it is a contingent matter that there are any stories. Lots of *A*-objects might have been stories, however. To say this is to say that they encode just vacuous properties and that POSSIBLY there exists an object which authors it.

Stories do not have to be consistent, nor do they have to be maximal. But stories and worlds do have something in common – they encode only vacuous properties. It is therefore appropriate to use our defined operator "Σ" to talk derivatively about the propositions the stories encode. In fact, if z is a story, then we may utilize "Σ_z" as our translation for the English prefix "according to (in) the story z". So when z is a story, "$\Sigma_z F^0$" says that F^0 is true according to z. This allows us to prove an interesting consequence of $\boldsymbol{D}_{34}$ which helps us to identify a given story: a story z is just that abstract object which encodes exactly the properties F which are constructed out of propositions true according to the story. That is,

THEOREM ("STORIES"): $(z)(Story(z) \rightarrow z = (\iota z')(F)(z'F \equiv (\exists F^0)(\Sigma_z F^0 \,\&\, F = [\lambda y F^0])))$[17].

For example, Little Red Riding Hood is a story, so it is that abstract object which encodes exactly the vacuous properties constructed out of propositions true according to Little Red Riding Hood. Although this is not a definition of "Little Red Riding Hood", we can identify this story in so far as we have a good pretheoretical idea about which propositions are true according to it. Fortunately, the data begins where our suggestion ends, for we suppose that the data are intuitively true English sentences of the form "according to the story,...". For example,[18]

(1) According to *The Tempest*, Prospero had a daughter.

(2) According to *The Iliad*, Achilles fought Hector.

(3) In *The Brothers Karamazov*, everyone that met Alyosha loved him.

(4) In *The Clouds*, Strepsiades converses with Socrates.

(5) In Joyce's *Ulysses*, Bloom journeys through Dublin.

Thus, STORIES helps us to understand which A-objects might be denoted by the underlined terms in the above sentences. We next try to identify the denotations of some of the other terms.

We can say what it is to be a character of a story. Let us use "s" variables as restricted variables ranging over stories:

$\boldsymbol{D}_{35}$ *x is a character of s* ("*Char*(x,s)") $=_{df}(\exists F)\Sigma_s Fx$.

That is, the characters of a story are the objects which exemplify properties according to it. As we noted previously, the characters of a story are any story objects, not just real or imaginary persons or animals. Note also that this definition allows existing objects to be characters of stories – we can tell stories (true or false) about existing objects, just as we can about non-existent ones.[19]

Of the non-existent characters in a given story, some will have originated entirely from that story. We call these the "native" characters, and they are to be distinguished from the other non-existent characters which may have been borrowed or imported from other stories. But the non-native non-existent characters are nevertheless "fictional", since, presumably, they are native to (originate from) some other story.

We may define the notions of being native and being fictional by utilizing a higher order primitive relation – one which could be analyzed in the context of some other work. This is the relation that two propositions F^0 and G^0 bear to one another just in case F^0 ***occurs*** (***obtains, takes place***)

before G^0. We shall represent the fact that F^0 occurs before G^0 as "$F^0 < G^0$". This relation helps us to be more specific about what it is to originate in a story:

$\boldsymbol{D}_{36}$ *x originates in s* ("$Origin(x,s)$") $=_{df}$
$Char(x,s)\ \&\ A!x\ \&\ (y)(y')(s')(Ays\ \&\ Ay's'\ \&\ (Ay's' < Ays) \rightarrow {\sim} Char(x,s'))$.

That is, x originates in s iff x is an abstract object which is a character of s and which is not a character of any earlier story. We then define being native and being fictional as follows:

$\boldsymbol{D}_{37}$ *x is native to s* ("$Native(x,s)$") $=_{df} Origin(x,s)$.

$\boldsymbol{D}_{38}$ *x is fictional* ("$Fict(x)$") $=_{df} (\exists s) Native(x,s)$.

So fictional characters are native to (originate in) some story. Clearly, fictional characters may be characters of stories to which they are not native. Sherlock Holmes is not native to *The Seven Per Cent Solution.*[20] Nor is the monster Grendel, in John Gardner's recent account of the Beowulf legend from the monster's point of view (*Grendel*). For simplicity, we shall suppose that Achilles and Hector are native to *The Iliad*, even though they may instead be native to some earlier epic of which no copies have survived. Also, in what follows, we shall suppose that Prospero is native to *The Tempest*, Alyosha and Raskolnikov are native to *The Brothers Karamazov* and *Crime and Punishment*, respectively, Bloom is native to Joyce's *Ulysses*, and Gregor Samsa is native to Kafka's *Metamorphosis.*

It would be a philosophical achievement of great importance were someone to discover a way of identifying fictional characters in general. The best we can accomplish here is to present a means of identifying the characters native to a given story. The identifying properties of native characters are exactly the properties exemplified by that character in the story. So we may utilize the following axiom which identifies the native characters of a story as specific A-objects:[21]

AXIOM ("*N*-CHARACTERS"): $(x)(s)(Native(x,s) \rightarrow x = (\iota z)(F)(zF \equiv \Sigma_s Fx))$.

For example, since Prospero is native to *The Tempest*, Prospero is that abstract object which encodes exactly the properties Prospero exemplifies according to *The Tempest*. This tells us an important fact about the

Σ_s-operator and native characters – the Σ_s-operator "transforms" a property a native character exemplifies according to story s into one which the character encodes. That is, it is a theorem that:[22]

$$(x)(s)(Native(x, s) \rightarrow (F)(xF \equiv \Sigma_s Fx)).$$

So if according to the play, Prospero had a daughter, it follows that he encodes having a daughter.

This theorem assumes greater significance in the presence of the following axiom schema which also should govern the Σ_s-operator:

AXIOM(S) ("Σ_s-SUBSTITUTION"): where ϕ is any propositional formula in which there occurs an object term o for which x is substitutable, the following is an axiom: $(s)(\Sigma_s\phi \rightarrow \Sigma_s[\lambda x\, \phi_o^x]o)$.

For example, in the myth, Achilles fought Hector. It therefore follows from Σ_s-SUB both that in the myth, Achilles exemplifies the property of fighting Hector and that in the myth, Hector exemplifies the property of being fought by Achilles. From the supposition that Achilles and Hector are both native to the myth in question, we may deduce that they encode these properties, respectively, by N-CHARACTERS.[23]

With these definitions, axioms, and consequences, we can translate a wide variety of data. We begin with (1)–(5) above. The translation procedure is straightforward – since the Σ_s-operator is defined only on proposition terms, we translate the English "in the story" using the operator, and translate the rest of the sentence just as we would into an ordinary predicate calculus:

(1)′ $\Sigma_{Tempest}(\exists y)Dyp$

(2)′ $\Sigma_{Iliad}Fah$

(3)′ $\Sigma_{BK}(x)(Mxa \rightarrow Lxa)$

(4)′ $\Sigma_{Clouds}Cs_1s_2$

(5)′ $\Sigma_{Ulysses}Jbd.$

There is an interesting class of sentences relevantly similar to (1) which we should discuss briefly. These true sentences begin with the story prefix and involve the predicative copula "is". For example, (6) and (7):

(6) According to *Crime and Punishment*, Raskolnikov is a student.

(7) In the Conan Doyle novels, Holmes is a detective.

Frequently, there are contexts in which it is acceptable to drop the story prefix and just use the remainder of the sentence. We can think of the resulting sentences "Raskolnikov is a student", "Holmes is a detective", as true if we suppose that the English copula "is" can be read as "encodes". We can therefore assimilate another phenomenon which is compatible with our earlier discovery about the ambiguity of "is".[24] In fact, it should be clear that all data like (1)–(5) (and not just those involving the copula "is") are subject to a structural ambiguity involving predication itself. In the context of the story operator, that data must be translated as an exemplification predication. Outside such a context, they must be understood as encoding predications.

I think we can partially accommodate the views of philosophers who object to (4)′ and (5)′ by arguing that the real Socrates and the real Dublin are not characters of *The Clouds* and *Ulysses*, respectively. We do this by supposing, instead, that the objects known as "the Socrates of *The Clouds*", and "the Dublin of *Ulysses*", are the relevant characters of these stories. We could suppose that these latter objects were native to these stories and use *N*-CHARACTERS to identify them. Such a procedure could be broadened to identify all non-native fictional characters. For example, we could say that the Sherlock Holmes of *The Seven Per Cent Solution* is native to that work, even though Sherlock Holmes is not.

The problem with this procedure is that one is forced to say something about the relationships between the real Socrates and the Socrates of *The Clouds*, between the Sherlock Holmes native to the Conan Doyle novels and the Sherlock Holmes native to *The Seven Per Cent Solution*, etc. This is no easy task. Clearly, the notion of weak correlation or embedding would not be of much help – the Socrates of *The Clouds* exemplifies-according-to-*The Clouds* (and consequently, encodes) properties not exemplified by the real Socrates. A full discussion of the host of problems which arise here would take us too far afield. Much further investigation is warranted before this procedure is to be adopted. Let us then turn to the next group of data.

(8) Santa Claus does not exist.

(9) Santa Claus might have existed.

(10) Franz Kafka wrote about Gregor Samsa.

(11) Some Greeks worshipped Dionysus.

(12) Prospero is a character of *The Tempest*.

(13) Raskolnikov is a fictional student.

Now it seems to me that there are two important readings for (8):

(8)′ $\sim E!sc$

(8″) $\sim E!!sc$

(8′) is provable, once we have used *N*-CHARACTERS to identify Santa Claus. But if we symbolize the English word "exists" as we have done in (8′) as "*E*!", then (8) turns out to be a necessary truth, since *A*-objects necessarily fail to exist. This conflicts with (9), however. (8) seems to have a reading on which it is not necessary. We have captured this reading of (8) with (8″). (8″) asserts that no existing object exemplifies all the properties Santa Clause encodes (recall $\boldsymbol{D}_{10}$, Chapter II, Section 3, and the remarks at the very end of Chapter III, Section 4). This clearly is a contingent truth.

This second reading of (8) is important for our understanding of (9). We cannot represent (9) as "$\Diamond E!sc$" for its negation is provable from *N*-CHARACTERS and the assumption that Santa Claus is a native character. We capture the truth embedded in (9) as (9′):

(9′) $\Diamond E!!sc$

Surely in some possible world, there exists an object which exemplifies all the properties Santa Claus exemplifies in the legend (let us assume Santa Claus does not exemplify incompatible properties in the legend). These remarks about the proper translations of (8) and (9) and the consequences thereof apply to all other data similar to (8) and (9) which involve other names of native characters described by consistent stories.

We translate (10) and (11) using exemplification formulas because they involve extranuclear properties which *A*-objects may exemplify.

(10′) Wks

(11′) $(\exists y)(Gy \,\&\, Wyd)$.

Being written about and being worshipped are extranuclear properties (or so I am supposing). They were not ascribed to (exemplified by) Samsa and Dionysus in the relevant stories.

Given our work above, (12) should be translated as:

(12′) *Char*(*Prospero*, *The Tempest*).

However, (13) is a more subtle case. Being fictional is a notion we have defined – it may not be a property ("$[\lambda x\ \text{Fictional}(x)]$" is ill-formed). But being a student is a property that Raskolnikov encodes, since he is native to *Crime and Punishment* and exemplifies that property in the novel. Consequently, we may define:

$\boldsymbol{D}_{39}$ *x is a fictional student* ("*F-student* (*x*)") $=_{df}$ $(\exists s)(Native(x,s) \;\&\; \Sigma_s Sx)$,

where "*S*" denotes being a student. Then from the assumptions that Raskolnikov is native to *Crime and Punishment* and that he is a student according to that story, we have (13′) as a consequence:

(13′) *F-student*(*Raskolnikov*).

In fact, we can generalize and suppose there is a whole group of notions, each one defined with respect to a given property *G*:

$\boldsymbol{D}_{40}$ *x is a fictional* $G =_{df} (\exists s)(Native(x,s) \;\&\; \Sigma_s Gx)$.

So Holmes is a fictional detective, Achilles is a fictional Greek warrior, etc., given the appropriate assumptions.[25]

Finally, we discuss definite descriptions. Consider (14) and (15):

(14) The detective who lived at 221 Baker St. in the Conan Doyle novels is more famous than any real detective.

(15) In *Crime and Punishment*, Porphyry arrested the student who killed an old moneylender.

It would be inappropriate to read the description in (14) as "the object which exemplifies detectivehood, exemplifies living at 221 Baker St., and exemplifies being a character of the Conan Doyle novels", since this description fails to denote. But we often use the description in (14) to refer to Holmes. The proper way to translate it is as "the object which according to the Conan Doyle novels exemplifies both detectivehood and living at 221 Baker St". Using "MFT" to abbreviate "more famous than", and other obvious abbreviations, we may read (14) as:

(14)′ $(y)(Dy \;\&\; E!y \rightarrow MFT(\iota x)\Sigma_{CD}(Dx \;\&\; Lx)y)$.

This says that every existing detective *y* is such that the object which according to the Conan Doyle novels exemplifies both detectivehood and living at 221 Baker St. is more famous than *y*.

A similar reading must be given to the definite description in (15). The following would be the WRONG symbolization of (15):

$$\Sigma_{CP}Ap(\iota x)(Sx \;\&\; (\exists y)(OMLy \;\&\; Kxy)).$$

The definite description fails to denote anything, even though it is entirely within the scope of the story operator. There may not be an object which exemplifies being a student and which killed an old moneylender. Or there may be two. But there is exactly one object which according to *Crime and Punishment* exemplifies being a student and killing an old moneylender. Consequently, (15) is properly read as (15)′:

(15)′ $\Sigma_{CP}Ap\,(\iota x)\Sigma_{CP}(Sx\ \&\ (\exists y)(OMLy\ \&\ Kxy))$.

When we read (hear) definite descriptions in the context of a story, there is an implicit understanding that the description denotes a character of the story. This implicit understanding is captured by placing the appropriate Σ-operator immediately after the iota-operator of the description. This guarantees that the description, should it denote, denotes a character of the story. (These remarks should also cast light on a very common kind of definite description used in natural language: "the person who allegedly…", "the man who, according to recent sources,…".)

To see this, consider the above example (15′). If we assume that Raskolnikov is the object which according to *Crime and Punishment* is a student who killed an old moneylender, we can show that Raskolnikov is a character of that story. So assume (16):

(16) $r = (\iota x)\Sigma_{CP}Sx\ \&\ (\exists y)(OMLy\ \&\ Kxy)$.

By DESCRIPTIONS, it follows that according to *Crime and Punishment*, Raskolnikov is a student who killed an old money lender, i.e.,

(17) $\Sigma_{CP}Sr\ \&\ (\exists y)(OMLy\ \&\ Kry)$.

By Σ_{CP}-SUB, it follows that Raskolnikov exemplifies being a student who killed an old moneylender, i.e.,

(18) $\Sigma_{CP}[\lambda x\, Sx\ \&\ (\exists y)(OMLy\ \&\ Kyx)]r$.

So there is a property which Raskolnikov exemplifies according to *Crime and Punishment*. By $\boldsymbol{D}_{31}$, Raskolnikov is a character of that story. So by placing the story operator immediately after the iota operator in the description, we guarantee that the object denoted, if there is one, is a character of the story.

Finally, note that (16) is a true identity statement. From (15′) and (16), it follows that according to *Crime and Punishment*, Porphyry arrested Raskolnikov, i.e.,

(19) $\Sigma_{CP}Apr$.

Consider next,

(20) Ponce de Leon searched for the fountain of youth.

(21) According to the myth, the fountain of youth is in Florida.

(22) The fountain of youth might exist.

Where "M" names the relevant myth, "S" denotes the searching relation, "L" denotes the being located in relation, "Y" names the property of being a fountain the waters of which confer everlasting youth, "p" denotes Ponce de Leon, and "f" denotes Florida, then we read (20) and (21) as follows:

(20′) $Sp(\iota x)\Sigma_M Yx.$

(21′) $\Sigma_M L(\iota x)\Sigma_M Yxf.$

These readings are straightforward, given our earlier discussion. Since (22) is true, we want to be sure not to capture it as an attribution of possible existence to an abstract object. Recall the discussion of (8) and (9). The English word "exists" as it occurs in (22) is not to be translated by our primitive notion of existence. Instead, (22) must be understood along the lines of (9) as (22′):

(22′) $\Diamond E!!(\iota x)\Sigma_M Yx$

The above results should establish at least a PRIMA FACIE case for thinking that stories and characters are abstract objects. The groundwork has been laid for further investigations which might fill in more details. In many ways, declarative discourse among persons is like storytelling. It might be worthwhile to regard a given discourse of an individual as a story. All of the names and descriptions represent characters in the story. This might facilitate suspension of belief when something false or suspicious sounding arises. Various "eyewitness" versions of what happened in a given situation constitute different stories. For each consistent story, worlds can be described in which there are existing objects which exemplify there all of the properties the characters of the story exemplify according to the story. It seems to me that there are clear possibilities for future research here.

5. MODALITY AND DESCRIPTIONS

We now examine another class of English sentences which seem to be true A PRIORI. They have the form "The $F_1, F_2, \ldots, F_n$ is G", where G is necessarily implied by one of the F_i and where there is not an object which (uniquely) exemplifies $F_1, F_2, \ldots, F_n$. Here are some examples:

(1) The set of all non-self-membered sets is a set.

(2) The even prime number greater than two is not odd.

(3) The existent golden mountain has a shape.

For considerations similar to those in Chapter II, Section 2, we translate the English definite descriptions as *A*-object descriptions. Except in these cases, we translate "the F_1, $F_2, \ldots, F_n$" as "$(\imath z)(G)(zG \equiv F_1 \Rightarrow G \vee F_2 \Rightarrow G \vee \ldots \vee F_n \Rightarrow G)$", where "$F \Rightarrow G$" means that necessarily, everything exemplifying F exemplifies G.

Consequently, the English descriptions in (1)–(3) are represented as follows, using obvious abbreviations:

(a) $(\imath z)(G)(zG \equiv [\lambda x\, Sx \;\&\; (y)(y \in x \equiv Sy \;\&\; y \notin y)] \Rightarrow G)$

(b) $(\imath z)(G)(zG \equiv [\lambda x\, Nx \;\&\; Ex \;\&\; Px \;\&\; x > 2] \Rightarrow G)$

(c) $(\imath z)(F)(zF \equiv E! \Rightarrow F \vee G \Rightarrow F \vee M \Rightarrow F)$.

In the metalanguage, we signify this reading of the definite article as "the$_{\mathscr{A}}$", and we assimilate the reading of the definite article proposed in Chapter II, Section 2 to this reading. Let us abbreviate (a)–(c) respectively as $(\imath z)\psi_1 - (\imath z)\psi_3$. By *A*-DESCRIPTIONS, it follows that any property satisfying the formula on the right of the biconditional in ψ is encoded by the object denoted by the entire description.

Take (a) for example. Since the property of being a set is necessarily implied by the property of being a set of non-self-membered sets, it follows that $(\imath z)\psi_1 S$. Our representations of (1)–(3) turn out to be theorems:

(1′) $(\imath z)\psi_1 S$

(2′) $(\imath z)\psi_2[\lambda x \sim Ox]$

(3′) $(\imath z)\psi_3 S$.

This reading of the English definite article has another important application. Philosophers since Russell have been puzzled by the following two arguments:

(I) (4) Necessarily, the teacher of Alexander is a teacher.
(5) Aristotle is the teacher of Alexander.

---

(6) Necessarily, Aristotle is a teacher.

(II) (7) Necessarily, nine is greater than seven.
(8) Nine is the number of planets.

∴ (9) Necessarily, the number of planets is greater than seven.

We seem to have conflicting intuitions about each of these arguments. On the one hand, they both appear to be valid, since they seem to be based on a simple application of the rule of identity elimination. On the other hand, in each argument, the premises appear to be true and the conclusions false.

Philosophers have explained the conflict in one of two ways, depending on whether the English descriptions are taken to be contextually defined (in the traditional Russellian manner) or taken to be complex terms constructed with a primitive operator "the". Let us look at the Russellian explanation first.

If we ignore the fact that "Aristotle" and "Alexander" are supposed to be abbreviated descriptions, then a classic Russellian explanation of the problem starts with the supposition that (4), (5), (8), and (9) have a complex, rather than simple, logical form. Contexts in which descriptions appear are systematically eliminated in favor of contexts in which existential and uniqueness clauses make explicit the information implicit in the description. And the present situation is further complicated by the fact that when there are modal operators around, there is both a way to eliminate the description so that the existential and uniqueness clauses appear before the operator (wide scope) and a way to eliminate the description so that these clauses appear after the operator (narrow scope). Consequently, (4) and (9) each get two readings, whereas (5) and (8) each get one. If we let "a" denote Aristotle, let $\phi_4 = \ulcorner Txr \urcorner$ (i.e. "x taught Alexander"), and let $\phi_5 = \ulcorner Nx\{u|u \text{ is a planet}\} \urcorner$ (i.e. "x numbers the set of planets"), then the Russellian readings of (4)–(9) are as follows:

(4.1) $(\exists x)(\phi_4 \;\&\; (y)(\phi_{4x}^{\,y} \rightarrow y = x) \;\&\; \Box Tx)$ (wide scope)

(4.2) $\Box(x)((\exists y)(\phi_{4x}^{\,y} \;\&\; (u)(\phi_{4x}^{\,u} \rightarrow u = y) \;\&\; y = x) \rightarrow Tx)$
(narrow scope)

(5.1) $(\exists x)(\phi_4 \;\&\; (y)(\phi_{4x}^{\,y} \rightarrow y = x) \;\&\; a = x)$

(6.1) $\Box Ta$

(7.1) $\Box 9 > 7$

(8.1) $(\exists x)(\phi_5 \;\&\; (y)(\phi_{5x}^{\,y} \rightarrow y = x) \;\&\; 9 = x)$

$$(9.1) \qquad (\exists x)(\phi_5 \ \& \ (y)(\phi_{5x}^{\ y} \rightarrow y = x) \ \& \ \Box x > 7) \qquad \text{(wide scope)}$$

$$(9.2) \qquad \Box(x)((\exists y)(\phi_{5x}^{\ y} \ \& \ (v)(\phi_{5x}^{\ v} \rightarrow v = y) \ \& \ y = x) \rightarrow x > 7). \qquad \text{(narrow scope)}$$

So Russellians claim that there are really two arguments to consider, one invalid, the other valid, when accounting for the conflict of intuitions about Argument I. The invalid argument has (4.2) and (5.1) as premises, which assert, respectively, that in every possible world, if there is a unique teacher of Alexander there, then whoever it is is a teacher, and that there is a unique teacher of Alexander who happens to be Aristotle. These are both true, but they do not jointly imply the falsehood (6.1) that Aristotle was a teacher in every possible world. The valid argument has (4.1) and (5.1) as premises, and they jointly imply the falsehood (6.1). But there is no cause for alarm because (4.1) is false, since the person who in fact taught Alexander did not teach Alexander in every world.

Both of the arguments which need to be considered in the case of argument (II) have true premises, since both (7.1) and (8.1) are true. But (7.1) and (8.1) jointly imply the truth (9.1), and do not imply the falsehood (9.2). From the facts that necessarily nine is greater than seven and that there's a unique object which numbers the set of planets and which happens to be nine, it does not follow that in every world, if there is a unique object there which numbers the planets in that world, then it is greater than seven.

The Russellian explanation of the apparent validity of Arguments (I) and (II) clearly works. The only trouble with it is that it doesn't preserve intuitions some of us may share about the logical form of (4), (5), (8), and (9). Some of us may share the intuition that the logical form of the sentences which follow the adverb "necessarily" in (4) and (9) is rather simple. These sentences seem to be atomic sentences with a complex subject term. And some of us may share the intuition that (5) and (8) are simple identity statements, constructed out of a primitive name, a complex term, and the "is" of identity. These intuitions are not preserved when (4), (5), (8), and (9) are represented in the Russellian fashion.

Philosophers who take these intuitions seriously will locate the source of trouble in the above arguments somewhere else. These philosophers will take the English definite article "the" as a primitive, represent it with the Greek letter iota, and construct complex terms like "$(\iota x)\phi$" to represent the English descriptions. Since it is taken as data that (4) has at least one true reading and that (9) has at least one false reading, these descriptions must be interpreted as non-rigid designators. There just doesn't seem to be any alternative for preserving the truth values of (4) and (9) in exemplification logic with primitive descriptions, since the use of rigid descriptions to represent the English descriptions in (4) and (9) would yield readings on which (4) was false and (9) was true. So (4)–(9) are customarily translated as follows, where the descriptions are not rigid:

(4a) $\Box T(\iota x)\phi_4$

(5a) $a = (\iota x)\phi_4$

(6a) $\Box Ta$

(7a) $\Box 9 > 7$

(8a) $9 = (\iota x)\phi_5$

(9a) $\Box(\iota x)\phi_5 > 7.$

On these representations, (4) and (9) are not ambiguous and there is no question of wide and narrow scope for the descriptions – all the descriptions are within the scope of the modal operator. And this fact, it is claimed, is just what is causing the trouble. The sentences "$T(\iota x)\phi_4$" and "$(\iota x)\phi_5 > 7$" may denote different propositions from world to world because the descriptions in them may denote different objects from world to world. So if one of the terms in a contingently true identity statement like (5a) or (8a) can change denotation from world to world, it is illegitimate to use the rule of identity elimination to substitute one of the terms for the other inside a modal context. Identity elimination is a rule which will preserve truth in modal contexts only if either the identity statement itself is a necessary truth or both of the terms in the identity statement are rigid designators (in which case, the identity statement will again be necessary). So the second standard kind of explanation about the tension we feel with respect to Arguments (I) and (II) is that they are, in fact, invalid. A properly stated rule of identity elimination makes the inferences illegitimate.

This latter explanation also clearly works. However, it fails to preserve the intuition some of us may share that Arguments (I) and (II) are simple valid arguments. And some of us may share the intution that there is a way to resolve the conflict without having to place "inelegant" restrictions on the rule of identity elimination. A natural suggestion to make has generally run up against a difficulty. The natural suggestion is to represent the English descriptions in these arguments as rigid designators. Then (4)–(9) get translated just as (4a)–(9a) above, except the descriptions in (4a), (5a), (8a), and (9a) denote in a given world, whoever it is that uniquely satisfies the description in the actual world. On this interpretation of the description, (4a) and (5a) logically imply (6a), because if in every possible world, the person who taught Alexander in the actual world is a teacher at that possible world, and if Aristotle is the teacher of Alexander in the actual world, then it must surely follow that Aristotle was a teacher in every possible world. And in a similar manner, (7a) and (8a) logically imply (9a), on this reading of the description. The rule of identity elimination preserves truth no matter what the context.

As we previously noted, however, the difficulty with this proposal is that it does not square with our intuitions that (4) seems to be true and (9) seems to be false. Construing the descriptions in (4a) and (9a) as rigid leaves (4a) false and (9a) true. There is no reason to think that the object which taught Alexander in the actual world (i.e. Aristotle) was a teacher in every possible world. Nor is there reason to think that there is a world where the object which in fact numbers the planets (i.e., the number nine) is less than seven. So we have not accounted for the intuitive truth values of (4) and (9). In fact, these results have lead many philosphers to conclude both that we cannot use rigid descriptions to represent these English descriptions and that, therefore, we need to resolve our conflict of intuitions about Argument (I) and (II) in one of the two ways outlined above.

These conclusions are not warranted however. There is a solution which both allows us to use rigid descriptions to preserve the intuitive truth values of (4) and (9), and allows us to resolve the conflict of intuitions over Arguments (I) and (II). The solution is based on suppositions for which we have found considerable evidence. They are that the English "is" sometimes should be read as "encodes" and the English "the" sometimes should be read as "the$_{\mathscr{A}}$". It is then straightforward to claim: that (4) and (9) are ambiguous, that the reading on which (4) comes out true is that necessarily the$_{\mathscr{A}}$ teacher of Alexander encodes being a teacher, and that the reading on which (9) comes out false is that necessarily, the$_{\mathscr{A}}$ number of planets encodes being greater than seven. To be precise, let us translate these new readings for the English descriptions in (4) and (9) as (d) and (e), respectively:

(d) $(\iota z)(G)(zG \equiv [\lambda x\, \phi_4] \Rightarrow G)$

(e) $(\iota z)(G)(zG \equiv [\lambda x\, \phi_5] \Rightarrow G)$.

Now, let us abbreviate the descriptions in (d) and (e) as "$(\iota z)\psi_4$" and "$(\iota z)\psi_5$", respectively. We then propose the following translations for (4)–(9), where all the descriptions are rigid:

(4a) $\Box\, T(\iota x)\phi_4$

(4b) $\Box\, (\iota z)\psi_4 T$

(5a) $a =_E (\iota x)\phi_4$

(6a) $\Box\, Ta$

(7a) $\Box\, 9 > 7$

(8a) $9 =_E (\iota x)\phi_5$

(9a) $\Box\, (\iota x)\phi_5 > 7$

(9b) $\Box\, (\iota z)\psi_5 [\lambda x\, x > 7]$.

These representations preserve all of the following intuitions: (i) that (4) seems to say something true (just consider (4b) and the fact that the property of being a teacher is implied by the property of being a teacher of Alexander); (ii) that (9) seems to say something false (just consider (9b) and the fact that the property of being greater than seven is not necessarily implied by the property of being something which numbers the set of planets); (iii) that the sentences following the adverb "necessarily" in (4) and (9) have a simple logical form (just consider the fact that (4a), (4b), (9a), and (9b) all involve atomic formulas); (iv) that Arguments (I) and (II) are simple valid arguments based on the simple rule of identity elimination (just consider that identity elimination works unrestricted in all contexts, and legitimately takes us from (4a) and (5a) to (6a), and from (7a) and (8a) to (9a)).

Of course the reader may not share these intuitions and consequently may not be moved by these results. Or the reader may have further intuitions about Arguments (I) and (II) which have not been preserved. If either of these are the case, then we may at least claim to have shown that we are not forced to accept the two traditional ways of explaining certain conflicting intuitions we have about Arguments (I) and (II).

An obvious plan for further investigation is to try to find data which involve descriptions and which force us to use non-rigid descriptions to preserve the intuitive truth values. Rigid descriptions clearly prove useful for understanding why (10) seems true:

(10) The inventor of bifocals might not have invented bifocals.

A straightforward translation of (10) can capture it as an atomic sentence. Consider (10′):

(10′) $[\lambda x \Diamond {\sim} Ix](\iota y)Iy.$

Since "$(\iota y)Iy$" has a denotation, we may use λE to prove (11) from (10′):

(11) $\Diamond {\sim} I(\iota y)Iy.$

(11) would be false were the description non-rigid, since in no possible world would the person that invented bifocals in that world fail to invent bifocals. So we're forced to use a rigid description to capture the truth in (10) (though if we are prepared to give up the intuition that it has a simple logical form, we could use a Russellian elimination in which the description gets wide scope to get a true reading). The question is though, will rigid descriptions and the logic of encoding always suffice?

CHAPTER V

THE TYPED THEORY OF ABSTRACT OBJECTS

The typed version of our theory commits us not only to abstract objects, but also to abstract properties, abstract relations, abstract properties of properties, abstract properties of relations, etc. We can use these entities to model impossible relations, like the symmetrical, non-symmetrical relation, and fictional relations, like simultaneity.[1] However, the primary motivation for developing the typed theory is to account for the data concerning the propositional attitudes and to model the fictional relations of mathematics.

The verbs of propositional attitude (e.g., believes, knows, desires, hopes, expects, discovers, etc.), often combine with the word "that" and an English sentence to produce logically problematic predicates like "believes that Cicero was a Roman", and "hopes that Kennedy is elected President." Frege noticed that terms (simple, complex names) inside these propositional attitude constructions exhibit rather strange behavior. In particular, Frege noticed that from the fact that someone believes that $\ldots\tau_1\ldots$, it doesn't follow that they believe that $\ldots\tau_2\ldots$, even when $\tau_1 = \tau_2$ (where $\ldots\tau_1\ldots$ is any English sentence in which term τ_1 occurs, and $\ldots\tau_2\ldots$ is the result of replacing one occurrence of τ_1 with τ_2).[2] For example, each of the following triads of English sentences is consistent:

(1) S believes that Cicero was a Roman.
(2) S does not believe that Tully was a Roman.
(3) Cicero is Tully.

(4) S believes that Socrates was the teacher of Plato.
(5) S does not believe that the son of Phaenarete was the teacher of Plato.
(6) Socrates is the son of Phaenarete.

(7) S believes that x is French fire engine blue.
(8) S does not believe that x is Crayola crayon blue.
(9) French fire engine blue just is Crayola crayon blue.[3]

It seems that the law of identity elimination ($=$E) does not preserve truth when applied to terms in propositional attitude contexts, and this constitutes the problem of "the logically deviant behavior of terms in intermediate contexts".

If in a given case, the law of identity elimination appears to fail, philosophers call the belief (context) DE DICTO, and distinguish it from a belief (context) DE RE, in which identity elimination preserves truth. When S's belief is DE RE, it does follow from the facts that S believes that $\ldots\tau_1\ldots$ and $\tau_1 = \tau_2$, that S believes that $\ldots.\tau_2\ldots.$

To account for this phenomenon of DE DICTO propositional attitudes, Frege theorized that there must be distinct entities, "senses", associated with the terms τ_1 and τ_2. These entities lend the terms with which they are associated information, or cognitive, value by serving somehow to RE-present the object or relation denoted by the term. This "mode of presentation" embodied by the sense of the term stores information about the denotation of the term, assuming it to have one. And it is the sense of the term which the term denotes when it is situated in a DE DICTO context. Frege would argue that identity elimination is a perfectly good rule of inference; it is just that English terms are ambiguous, and have different denotations when they are in and out of DE DICTO contexts. Identity elimination preserves truth when you substitute terms which have the same denotation.

Using the theory we have so far, we could construe the senses of English names and descriptions which denote objects as abstract objects. An association of abstract objects with English terms would allow us to picture how a given term had "information" or "cognitive" value. Abstract objects could "RE-present" an object denoted by a term by encoding properties the object exemplified. They could serve to store information by encoding many such properties. Finally, they could serve as the denotation of the term when the term is located inside DE DICTO contexts.

Such an association between terms denoting objects and abstract objects is one of the most important features of the language developed in this chapter. We use this language to translate data similar to (1)–(6) in Section 1 of Chapter VI. However, (7)–(8)–(9) constitute an example of the DE DICTO phenomenon with respect to English names which denote relations. "French fire engine blue" and "Crayola crayon blue" are names of certain properties – properties which we could suppose are identical. In order to account for the logically deviant behavior of these names, we associate with them abstract properties – properties which encode properties of properties. These abstract properties can lend property names their information value – they could store information about the properties denoted by such names by encoding properties of them. And these abstract properties can serve as the denotation of these names when the name is located in a DE DICTO context.

Similarly with English names which denote relations – we utilize abstract relations, relations which encode properties of relations, to serve as their sense. A completely general account of the senses of names of relations in the type hierarchy requires that we have abstract entities at each type which encode properties of the entities of that type. This is by far one of the most interesting applications of the typed version of our theory.

In what follows, we shall use the word "object" in a new manner. The things which we have been calling "objects" will now be called "individuals". We shall now use the term "object" to discuss any kind of entity whatsoever – existing and abstract individuals, existing and abstract properties and relations, existing and abstract properties of properties (relations), etc. Thus, we call the developments in the next few pages "the typed theory of abstract objects", and we affectionately refer to it as "metaphysical hyperspace".

1. THE LANGUAGE

We first recursively define the set of types.[4] For our purposes, we may think of types as symbols which serve to simultaneously categorize both the terms of the language and the entities denoted by those terms. The set of types, then, is the smallest set, TYPE, which satisfies the following conditions:

(1) "i"$\in$TYPE.

(2) "p"$\in$TYPE.

(3) Whenever $t_1,\ldots,t_n \in$TYPE, then $\ulcorner (t_1,\ldots,t_n)/p \urcorner \in$TYPE.

Intuitively, "i" is the type of individuals and "p" is the type of propositions. "$(t_1,\ldots,t_n)/p$" is the type of relations whose arguments have types $t_1,\ldots,t_n$, respectively. In what follows, we drop the quotation marks around the type symbols. The properties (and the expressions which named them) which we used in Chapters I–IV were of type i/p. The relations were of type $(i,\ldots,i)/p$. But now we have an infinitely branching hierarchy.

A. PRIMITIVE TERMS

Officially, we use $a_1^t, a_2^t, \ldots$ as names, and $x_1^t, x_2^t, \ldots$ as variables for objects of each type t. These are the only primitive terms of the language. However, the following conventions shall hold. Whenever $a, b, c, \ldots$ and $x, y, z, \ldots$ appear without typescripts, they denote (range over) individuals (unless their first occurrence in a formula has a typescript, in which case it shall be understood, if they appear later in the formula without typescripts, that the typescripts have been omitted for convenience). Also, we use $P^{(t_1,\ldots,t_n)/p}, Q^{(t_1,\ldots,t_n)/p}, \ldots$ and $F^{(t_1,\ldots,t_n)/p}, G^{(t_1,\ldots,t_n)/p}, \ldots$ as names and variables for objects of relational types $(t_1, \ldots, t_n)/p$. And we use $P^p, Q^p, \ldots$ and $F^p, G^p, \ldots$ as names and variables for objects of type p.

It will be convenient to distinguish certain names for special purposes. We use $E!^{t/p}$ as the existence predicate for objects of type t. We use $=_{E^t}$ as the E-identity predicate for objects of type t. We use $Tr^{p/p}$ as the explicit truth predicate for propositions. We use $Ex^{((t_1,\ldots,t_n)/p,t_1,\ldots,t_n)/p}$ as the explicit exemplification predicate, for all types $t_1, \ldots, t_n$. We use $B_1^{(i,p)/p}, B_2^{(i,p)/p}, \ldots$ to translate the verbs of propositional attitudes. Finally, we use $R^{(t,t,i)/p}$ as the representation predicate – some objects of type t will represent other objects of type t for an individual of type i.

In addition to these terms, we utilize our usual list of logical and grammatical symbols: connectives: $\sim, \rightarrow$; quantifier: $\forall$; lambda: λ; iota: ι; box: $\Box$; and parentheses and brackets: (,), [,]. We add to this list a one-place sentential operator: *that*-.

B. FORMULAS AND TERMS[5]

We simultaneously define **(propositional) formula** and **term of type** t. The definition has eight clauses and is rather complex. We reserve the extended comments and the examples until after the definition.

(1) All primitive terms of type t are terms of type t.

(2) Atomic: If τ is a term of type p, then τ is a (propositional) formula.

(3) Atomic exemplification: If ρ is a term of type $(t_1, \ldots, t_n)/p$, and $\tau_1, \ldots, \tau_n$ are terms with types $t_1, \ldots, t_n$, respectively, then $\rho\tau_1, \ldots, \tau_n$ is a (propositional) formula.

In atomic exemplification formulas, we call ρ the **initial term** and $\tau_1, \ldots, \tau_n$

are called **argument** terms. Primitive terms of type p will also be called **initial terms. Initial variables** are variables which are initial terms

(4) Atomic encoding: If ρ is a term of type t/p and τ is a term of type t, then $\tau\rho$ is a formula.

(5) Molecular, Quantified, and Modal: If ϕ, ψ are any (propositional) formulas and α is any variable of type t (which is not an initial variable somewhere in ϕ), then $(\sim\phi)$, $(\phi \rightarrow \psi)$, $(\forall\alpha)\phi$, and $(\Box\phi)$ are (propositional) formulas.

(6) Complex higher order terms: If ϕ is any propositional formula and $\alpha_1, \ldots, \alpha_n$ are any variables with types $t_1, \ldots, t_n$, respectively, such that none of the α_i's are initial variables somewhere in ϕ, then $[\lambda\alpha_1 \ldots \alpha_n \phi]$ is a term of type $(t_1, \ldots, t_n)/p$ and both ϕ itself and *that*-ϕ are terms of type p (it will sometimes be convenient to regard ϕ as the degenerate λ-expression $[\lambda\,\phi]$).

(7) Sense terms: If κ^t is any primitive name of type t, and σ is any primitive term of type i, $\underline{\kappa}^t_\sigma$ is a term of type t.

(8) (Sense) Descriptions: If ϕ is any (propositional) formula with one free variable x of type t, then $(\iota x^t)\phi$ $(\underline{(\iota x)\phi})$ is a term of type t.

In the usual manner, we define:

D_1 x^t *is abstract*$^{t/p}$ ("$A!^{t/p}x^t$") $=_{df} [\lambda y^t\, \Box \sim E!^{t/p}y]x^t$.

We shall use z^t-variables to range over abstract objects of type t.

By inserting the parenthetical remarks in clauses 2, 3, 4, and 5, we obtain a definition of propositional formula. Essentially, a formula ϕ is propositional iff ϕ has no encoding subformulas and none of the initial variables appearing in ϕ are bound by a quantifier. Clearly, $(\exists G^{i/p})(x^iG\ \&\ \sim Gx)$ and $(G^{i/p})(x^iG \rightarrow Gx)$ fail both of these restrictions. However, $(\forall F^{i/p})Fx^i$, $(\exists F^{i/p})(G^{i/p}x^i\ \&\ \sim Fx^i)$, $(\forall F^p)(F^p \vee \sim F^p)$, and $(\exists G^p)(F^{i/p}x^i\ \&\ \sim G^p)$ all fail the second restriction. As we noted in earlier chapters, the second restriction allows us to simplify the semantics. But now that we have a type theory, there are other ways to express the propositions these latter four formulas seem to express. For example, we could use the explicit exemplification predicate of type $(i/p,i)/p$ and suppose that $(\forall F^{i/p})Fx^i$ was simply an abbreviation for $(\forall F^{i/p})ExFx^i$. This latter formula is propositional, and says that all properties of individuals $F^{i/p}$ are such that they bear the exemplification relation of type $(i/p, i)/p$ to x^i. Also, we could use the explicit truth predicate and suppose that $(\forall F^p)(F^p \vee \sim F^p)$ abbreviated $(\forall F^p)(TrF^p \vee \sim TrF^p)$. The latter formula is propositional and says that all propositions F^p are such that either F^p exemplifies the property of being true or fails to exemplify this property. In general, we can always reconstitute a propositional formula from a formula which fails just the second restriction on propositional formulas.[6] This lessens the significance of the second restriction.

In clause 6, there is an additional restriction placed on the formulas ϕ which may appear behind a λ: variables bound by the λ must not appear as an initial variable somewhere in ϕ. Techniques similar to the above allow us to reconstruct a propositional formula satisfying this restriction from one which fails it. For example, $[\lambda F^{i/p} Fx^i]$, an expression which fails this restriction, could abbreviate $[\lambda F^{i/p} ExFx^i]$, where the Ex predicate is of the appropriate type. $[\lambda F^p F^p \rightarrow F^p]$ could abbreviate $[\lambda F^p TrF^p \rightarrow TrF^p]$; the latter denotes the reflection of the conditionalization of the property of being true with itself. Abbreviational procedures such as these lessen the significance of the restriction in clause 6. In fact, these abbreviational techniques allow us to construe any λ-expression $[\lambda \alpha_1 \ldots \alpha_n \phi]$ in which ϕ lacks encoding subformulas either as a well-formed λ-expression or as an abbreviation of a well-formed λ-expression. But they do not allow us to construe $[\lambda x^i(\exists F^{i/p})(xF \,\&\, \sim Fx)]$ or $[\lambda x^i(G^{i/p})(xG \rightarrow Gx)]$ as abbreviations of well-formed λ-expressions.

Clause 7 of the above definition gives us a means for denoting the abstract object an individual associates with a given English proper name as its sense. We suppose that the sense of a name varies from person to person (see Chapter VI, Section 1). For example "Socrates" and "Frege" are names of type *i*, so "$\underline{\text{Socrates}}_{\text{Frege}}$" is a sense term of type *i*. It shall denote the abstract individual which serves as the sense of the name "Socrates" with respect to Frege. "French fire engine blue" is a name of type i/p (since it names a property of individuals). So "$\underline{\text{French fire engine blue}}_{\text{Frege}}$" is a sense term of type i/p and will denote the abstract property which serves to represent the property of being French fire engine blue to Frege. This abstract property encodes properties of type $(i/p)/p$, i.e., properties of i/p-properties.

Clause 8 of our definition simultaneously gives us both descriptions and sense descriptions. Where "T" denotes the $(i, i)/p$-relation of teaching, and "p" denotes Plato, $(\iota x)Txp$ reads "the teacher of Plato". Where "C" denotes the $(i/p)/p$-property of being a color, "P" denotes the preference relation of type $(i,i/p,i/p)/p$, and "m" denotes Mary, $(\iota x^{i/p})(Cx \,\&\, \sim(\exists y)(Cy \,\&\, Pmyx))$ might read: the color Mary prefers to all others (i.e. "Mary's favorite color").

If the formula ϕ used in constructing the description is propositional, then $\underline{(\iota x^t)\phi}$ is called a sense description. They will help us to model the senses of English definite descriptions. $\underline{(\iota x^t)\phi}$ shall end up denoting the abstract object of type t which encodes just the property $[\lambda x^t \phi \,\&\, (y)(\phi^y_x \rightarrow y =_E x)]$ ("being the unique ϕ"). For example, $\underline{(\iota x^i)Txp}$ shall denote the abstract individual which encodes just the property of being the teacher of Plato. When we concern ourselves specifically with the fact that the English description "the teacher of Plato" exhibits logically deviant behavior inside DE DICTO attitude contexts, we shall translate the English as we normally would into the standard type theoretic language and then underline it. By doing so, we will have formed an expression of our language which denotes the sense of the English description.

Finally, we say that τ is a **term** iff there is a type t such that τ is a term of type t.

2. THE SEMANTICS

A. INTERPRETATIONS

An interpretation, $\mathscr{I}$, of our type theoretic language is any octuple, $\langle \mathscr{W}, \mathscr{w}_0, \mathscr{D}, \mathscr{ext}_{\mathscr{w}}, \mathscr{L}, \mathscr{ext}_{\mathscr{A}}, \mathscr{sen}, \mathscr{F} \rangle$, which meets the conditions described in this subsection. The first member of $\mathscr{I}$ is a non-empty class, $\mathscr{W}$, called the class of **possible worlds**. The second member of $\mathscr{I}$, $\mathscr{w}_0$, is a member of $\mathscr{W}$ and is called the **actual world**. The third member of $\mathscr{I}$, $\mathscr{D}$, is a non-empty class called the **domain of objects**. $\mathscr{D}$ is the union of a collection of non-empty, indexed, classes, i.e., $\mathscr{D} = \bigcup_{t \in \mathrm{TYPE}} \mathscr{D}_t$. Each class in the collection, $\mathscr{D}_t$, is called the **domain of objects of** type t. We call $\mathscr{D}_i$ the **domain of individuals,** $\mathscr{D}_p$ the **domain of propositions,** $\mathscr{D}_{t/p}$ the **domain of properties of type** t **objects,** $\mathscr{D}_{(t_1,\ldots,t_n)/p}$ the **domain of n-place relations among objects with types** $t_1,\ldots,t_n$, respectively. We use "$\mathscr{o}^t$" as a metalinguistic variable ranging over the objects in $\mathscr{D}_t$.

For convenience, we call the class of all objects with types not equal to i the class of **higher order objects** and we use "$\mathscr{R}$" to denote this domain. So $\mathscr{R} = \bigcup_{t \neq i} \mathscr{D}_t$. $\mathscr{R}$ is closed under all the logical functions specified in $\mathscr{L}$, the fifth member of an interpretation. $\mathscr{R}$ may be subdivided into domains of relational types $\mathscr{R}_{(t_1,\ldots,t_n)/p}$ and the domain of propositions $\mathscr{R}_p$. We use "$\imath^t$" as a metalinguistic variable ranging over the higher order objects of type t.

We also let "$\mathscr{A}_t$" denote the class of **abstract** objects of type t. $\mathscr{A}_t = \{o^t | (w)(o^t \notin ext_w(\mathscr{F}(E!^{t/p})))\}$, where ext_w and $\mathscr{F}$ are the fourth and eighth members of the interpretation, as defined below. We use "a^t" as metalinguistic variables ranging over the members of $\mathscr{A}_t$.

The fourth member of $\mathscr{I}$, ext_w, is a function defined on $\mathscr{R} \times \mathscr{W}$ and indexed to its second argument as follows:

(a) $ext_w: \mathscr{R}_{(t_1,\ldots,t_n)/p} \times \mathscr{W} \to \mathfrak{P}(\mathscr{D}_{t_1} \times \mathscr{D}_{t_2} \times \ldots \times \mathscr{D}_{t_n})$

(b) $ext_w: \mathscr{R}_p \times \mathscr{W} \to \{T, F\}$.

Thus, the ext_w function distributes an **exemplification extension** at each world to all the higher order objects.

The fifth member of $\mathscr{I}$, $\mathscr{L}$, is a class of logical functions with members: $\mathscr{PLUG}_j$, $\mathscr{UNIV}_j$, $\mathscr{CONV}_{j,k}$, $\mathscr{REFL}_{j,k}$, $\mathscr{VAC}_{j,t'}$, $\mathscr{COND}$, $\mathscr{NEG}$, and $\mathscr{NEC}$. These functions are defined as follows:

(a) $\mathscr{PLUG}_j$ is a function from $(\bigcup_{1 \le j \le n} \mathscr{R}_{(t_1,\ldots,t_j,\ldots,t_n)/p}) \times \mathscr{D}_{t_j}$ into $(\bigcup_{1 \le j \le n} \mathscr{R}_{(t_1,\ldots,t_{j-1},t_{j+1},\ldots,t_n)/p}) \cup \mathscr{R}_p$ subject to the following conditions:

(1) for $n > 1$, $ext_w(\mathscr{PLUG}_j(\imath^{(t_1,\ldots,t_n)/p}, o^{t_j})) = \{\langle o^{t_1},\ldots,o^{t_{j-1}},o^{t_{j+1}},\ldots,o^{t_n}\rangle | \langle o^{t_1},\ldots,o^{t_{j-1}},o^{t_j},o^{t_{j+1}},\ldots,o^{t_n}\rangle \in ext_w(\imath^{(t_1,\ldots,t_n)/p})\}$

(2) for $n = 1$, $ext_w(\mathscr{PLUG}_1(\imath^{t_1/p}, o^{t_1})) = \begin{cases} T \text{ iff } o^{t_1} \in ext_w(\imath^{t_1/p}) \\ F \text{ otherwise.} \end{cases}$

(b) $\mathscr{UNIV}_j$ is a function from $\bigcup_{1 \le j \le n} \mathscr{R}_{(t_1,\ldots,t_j,\ldots,t_n)/p}$ into $(\bigcup_{1 \le j \le n} \mathscr{R}_{(t_1,\ldots,t_{j-1},t_{j+1},\ldots,t_n)/p}) \cup \mathscr{R}_p$ subject to the conditions:

(1) for $n > 1$, $ext_w(\mathscr{UNIV}_j(\imath^{(t_1,\ldots,t_n)/p})) = \{\langle o^{t_1},\ldots,o^{t_{j-1}},o^{t_{j+1}},\ldots,o^{t_n}\rangle | (\forall o^{t_j})(\langle o^{t_1},\ldots,o^{t_{j-1}},o^{t_j},o^{t_{j+1}},\ldots,o^{t_n}\rangle \in ext_w(\imath^{(t_1,\ldots,t_n)/p}))\}$

(2) for $n = 1$, $ext_w(\mathscr{UNIV}_1(\imath^{t_1/p})) = \begin{cases} T \text{ iff } (\forall o^{t_1})(o^{t_1} \in ext_w(\imath^{t_1/p})) \\ F \text{ otherwise.} \end{cases}$

(c) $\mathscr{CONV}_{j,k}$ is a function from $\bigcup_{1 \le j < k \le n} \mathscr{R}_{(t_1,\ldots,t_j,\ldots,t_k,\ldots,t_n)/p}$ into $\bigcup_{1 \le j < k \le n} \mathscr{R}_{(t_1,\ldots,t_{j-1},t_k,t_{j+1},\ldots,t_{k-1},t_j,t_{k+1},\ldots,t_n)/p}$ subject to the

following condition:
$\mathscr{ext}_w(\mathscr{CONV}_{j,k}(\imath^{(t_1,\ldots,t_n)/p})) =$
$\{\langle o^{t_1},\ldots,o^{t_{j-1}},o^{t_k},o^{t_{j+1}},\ldots,o^{t_{k-1}},o^{t_j},o^{t_{k+1}},\ldots,o^{t_n}\rangle|$
$\langle o^{t_1},\ldots,o^{t_j},\ldots,o^{t_k},\ldots,o^{t_n}\rangle \in \mathscr{ext}_w(\imath^{(t_1,\ldots,t_n)/p})\}$.

(d) $\mathscr{REFL}_{j,k}$ is a function from $\bigcup_{1\le j<k\le n}\mathscr{R}_{(t_1,\ldots,t_j,\ldots,t_k,\ldots,t_n)/p}$ into $\bigcup_{1\le j<k\le n}\mathscr{R}_{(t_1,\ldots,t_j,\ldots,t_{k-1},t_{k+1},\ldots,t_n)/p}$ subject to the following condition:
$\mathscr{ext}_w(\mathscr{REFL}_{j,k}(\imath^{(t_1,\ldots,t_n)/p})) =$
$\{\langle o^{t_1},\ldots,o^{t_j},\ldots,o^{t_{k-1}},o^{t_{k+1}},\ldots,o^{t_n}\rangle|$
$\langle o^{t_1},\ldots,o^{t_j},\ldots,o^{t_k},\ldots,o^{t_n}\rangle \in \mathscr{ext}_w(\imath^{(t_1,\ldots,t_n)/p})$ and $o^{t_j} = o^{t_k}\}$.

(e) $\mathscr{VAC}_{j,t'}$ is a function from $(\bigcup_{1\le j\le n+1}\mathscr{R}_{(t_1,\ldots,t_j,\ldots,t_n)/p}) \cup \mathscr{R}_p$ into $(\bigcup_{1\le j\le n+1}\mathscr{R}_{(t_1,\ldots,t_{j-1},t',t_j,t_{j+1},\ldots,t_n)/p}) \cup \mathscr{R}_{t'/p}$ subject to the conditions:

(1) if $t = (t_1,\ldots,t_j,\ldots,t_n)/p$, then $1 \le j \le n+1$ and
$\mathscr{ext}_w(\mathscr{VAC}_{j,t'}(\imath^t)) = \{\langle o^{t_1},\ldots,o^{t_{j-1}},o^{t'},o^{t_j},o^{t_{j+1}},\ldots,o^{t_n}\rangle|$
$\langle o^{t_1},\ldots,o^{t_j},\ldots,o^{t_n}\rangle \in \mathscr{ext}_w(\imath^t)\}$

(2) if $t = p$, then $j = 1$ and $\mathscr{ext}_w(\mathscr{VAC}_{1,t'}(\imath^p)) =$
$\{o^{t'} | \mathscr{ext}_w(\imath^p) = T\}$.

(f) $\mathscr{COND}$ is a function from $\mathscr{R} \times \mathscr{R}$ into $\mathscr{R}$ subject to the following conditions:

(1) if $t = (t_1,\ldots,t_n)/p$ and $t' = (t'_1,\ldots,t'_m)/p$, then
$\mathscr{ext}_w(\mathscr{COND}(\imath^t,\jmath^{t'})) =$
$\{\langle o^{t_1},\ldots,o^{t_n},o^{t'_1},\ldots,o^{t'_m}\rangle | \langle o^{t_1},\ldots,o^{t_n}\rangle$
$\notin \mathscr{ext}_w(\imath^t)$ or $\langle o^{t'_1},\ldots,o^{t'_m}\rangle \in \mathscr{ext}_w(\jmath^{t'})\}$

(2) if $t = (t_1,\ldots,t_n)/p$ and $t' = p$, then
$\mathscr{ext}_w(\mathscr{COND}(\imath^t,\jmath^{t'})) =$
$\{\langle o^{t_1},\ldots,o^{t_n}\rangle | \langle o^{t_1},\ldots,o^{t_n}\rangle \notin \mathscr{ext}_w(\imath^t)$
or $\mathscr{ext}_w(\jmath^{t'}) = T\}$

(3) if $t = p$ and $t' = (t_1,\ldots,t_m)/p$, then $\mathscr{ext}_w(\mathscr{COND}(\imath^t,\jmath^{t'})) =$
$\{\langle o^{t_1},\ldots,o^{t_m}\rangle | \mathscr{ext}_w(\imath^t) = F$ or
$\langle o^{t_1},\ldots,o^{t_m}\rangle \in \mathscr{ext}_w(\jmath^{t'})\}$

(4) if $t = p$ and $t' = p$, then $\mathscr{ext}_w(\mathscr{COND}(\imath^t,\jmath^{t'})) =$
$\begin{cases} T \text{ iff } \mathscr{ext}_w(\imath^t) = F \text{ or } \mathscr{ext}_w(\jmath^{t'}) = T \\ F \text{ otherwise.} \end{cases}$

(g) $\mathcal{NEG}$ is a function from $\mathcal{R}$ into $\mathcal{R}$ subject to the conditions:

(1) if $t=(t_1,\ldots,t_n)/p$, then $ext_w(\mathcal{NEG}(r^t))=$
$\{\langle o^{t_1},\ldots,o^{t_n}\rangle \mid \langle o^{t_1},\ldots,o^{t_n}\rangle \notin ext_w(r^t)\}$

(2) if $t=p$, then $ext_w(\mathcal{NEG}(r^p))=\begin{cases} T \text{ iff } ext_w(r^p)=F \\ F \text{ otherwise.}\end{cases}$

(h) $\mathcal{NEC}$ is a function from $\mathcal{R}$ into $\mathcal{R}$ subject to the conditions:

(1) if $t=(t_1,\ldots,t_n)/p$, then $ext_w(\mathcal{NEC}(r^t))=$
$\{\langle o^{t_1},\ldots,o^{t_n}\rangle \mid (w')(\langle o^{t_1},\ldots,o^{t_n}\rangle \in ext_{w'}(r^t))\}$

(2) if $t=p$, then $ext_w(\mathcal{NEC}(r^p))=\begin{cases} T \text{ iff } (w')(ext_{w'}(r^p)=T) \\ F \text{ otherwise.}\end{cases}$

This completes the definitions of the logical functions. The sixth member of $\mathcal{I}$, $ext_{\mathcal{A}}$, is a function defined on $\bigcup_{t\in\text{TYPE}}\mathcal{R}_{t/p}$. For a given type t, $ext_{\mathcal{A}}$ maps $\mathcal{R}_{t/p}$ into $\mathfrak{P}(\mathcal{D}_t)$. $ext_{\mathcal{A}}$ assigns each higher order property of t-objects an **encoding extension** among these objects.

Let $\mathcal{N}_t$ be the set of primitive names of type t of our language. Then, the seventh member of $\mathcal{I}$ is the sense function, sen, which maps $\mathcal{D}_i \times \mathcal{N}_t$ into $\mathcal{A}_t$ (the set of abstract objects of type t). For convenience, we index the sen function to its first argument. Thus, for a given individual o, sen_o associates with a given name κ^t of type t an abstract object of type t. We call $sen_o(\kappa)$ **the sense of κ with respect to o**.[7] Intuitively, if "Socrates" is a name of type i, then $sen_{\mathcal{F}rege}$("Socrates") is the abstract individual which serves as the sense of the name "Socrates" with respect to Frege. We shall assign this object to the sense term "$\underline{\text{Socrates}}_{\text{Frege}}$". And we shall make it a logical truth that $\underline{\text{Socrates}}_{\text{Frege}}$ represents Socrates to Frege.[8] We shall sometimes index the sense function to the type of the name upon which it is operating. For example, $sen^{i/p}_{\mathcal{F}rege}$("French fire engine blue") is the abstract i/p-property which serves as the sense of "French fire engine blue" with respect to Frege.

The eighth member of $\mathcal{I}$ is a function, $\mathcal{F}$, defined on the primitive names AND on the closed sense terms of the language. For each name κ^t of type t, $\mathcal{F}(\kappa^t)\in\mathcal{D}_t$. For each closed sense term $\underline{\kappa}^t_\sigma$ of type t, $\mathcal{F}(\underline{\kappa}^t_\sigma)=sen_{\mathcal{F}(\sigma)}(\kappa^t)$. Recall that sense terms can have only primitive terms as subscript. So the closed sense terms will have only primitive names as subscripts.[9]

In addition, we place the following three restrictions on $\mathcal{F}$:

(1) $\mathit{ext}_{\omega}(\mathscr{F}(Ex^{((t_1,\ldots,t_n)/p,t_1,\ldots,t_n)/p})) = \{\langle \imath^{(t_1,\ldots,t_n)/p}, o^{t_1},\ldots,o^{t_n}\rangle \mid \langle o^{t_1},\ldots,o^{t_n}\rangle \in \mathit{ext}_{\omega}(\imath^{(t_1,\ldots,t_n)/p})\}$.

So $\mathscr{F}$ must assign to the explicit exemplification predicates relations with the "appropriate" extensions.

(2) $\mathit{ext}_{\omega}(\mathscr{F}(\mathrm{Tr})) = \{\imath^p \mid \mathit{ext}_{\omega}(\imath^p) = T\}$.

Here too, $\mathscr{F}$ must assign to the explicit truth predicate a property of propositions with the appropriate extension.

(3) $\mathit{ext}_{\omega}(\mathscr{F}(R^{(t,t,i)/p})) = \{\langle a^t, o^t, o^i\rangle \mid (\exists \underline{\kappa}_{\sigma})(\mathscr{F}(\underline{\kappa}_{\sigma}) = a^t \;\&\; \mathscr{F}(\kappa) = o^t \;\&\; F(\sigma) = o^i)\}$

Thus, "$R^{(t,t,i)/p}$" denotes a three place relation which objects a^t, o^t, and o^i bear to one another just in case there is some (closed) sense term $\underline{\kappa}_{\sigma}$ such that a^t is the sense of κ with respect to o^i and o^t is the denotation of κ. We say that a^t **represents** o^t **with respect to** o^i.

B. ASSIGNMENTS AND DENOTATIONS

For the most part, the definitions partitioning the λ-expressions are similar to those developed in Chapter III, Section 2, B. However, we need to type the added place in the definition of vacuous expansion. We also need to concern ourselves with argument variables (rather than the "object" variables of Chapter III) throughout these definitions.

If μ is an arbitrary λ-expression, $[\lambda\alpha_1\ldots\alpha_n\phi]$, μ is classified as follows:

(1) If $(\exists j)$ ($1 \leq j \leq n$ and α_j does not occur free in ϕ and t' is the type of α_j and j is the least such number), then μ is the **j, t′-vacuous expansion of** $[\lambda\alpha_1\ldots\alpha_{j-1}\alpha_{j+1}\ldots\alpha_n\phi]$.

(2) If μ is not a j,t'-vacuous expansion, then if $(\exists j)(1 \leq j \leq n$ and α_j is not the jth free argument variable in ϕ and j is the least such number), then where α_k is the jth free argument variable in ϕ, μ is the **j, kth-conversion of** $[\lambda\alpha_1\ldots\alpha_{j-1}\alpha_k\alpha_{j+1}\ldots\alpha_{k-1}\alpha_j\alpha_{k+1}\ldots\alpha_n\phi]$.

(3) If μ is neither of the above, then

(a) if $\phi = (\sim\psi)$, μ is the **negation of** $[\lambda\alpha_1\ldots\alpha_n\psi]$

(b) if $\phi = (\psi \rightarrow \chi)$, and ψ and χ have no free argument variables in common, then where $\alpha_1,\ldots,\alpha_m$ are the variables in ψ and $\alpha_{m+1},\ldots,\alpha_n$ are the variables in χ, μ is the **conditionalization of** $[\lambda\alpha_1\ldots\alpha_m\psi]$ **and** $[\lambda\alpha_{m+1}\ldots\alpha_n\chi]$.

(c) if $\phi = (\forall\beta)\psi$, and β is the jth free argument variable in ψ, then μ is the **jth-universalization of** $[\lambda\alpha_1 \ldots \alpha_{j-1}\beta\alpha_j\alpha_{j+1} \ldots \alpha_n\psi]$.

(d) if $\phi = (\Box\psi)$, then μ is the **necessitation of** $[\lambda\alpha_1 \ldots \alpha_n\psi]$.

(4) If μ is none of the above, then if $(\exists j)$ $(1 \leq j \leq n$ and α_j occurs free in more than one place in ϕ and j is the least such number), then where:

(a) m is the number of free argument variables between the first and second occurrences of α_j,

(b) ϕ' is the result of replacing the second occurrence of α_j with a new variable β (with the same type as α_j),

(c) $k = j + m + 1$, then
μ is the **j, kth-reflection of** $[\lambda\alpha_1 \ldots \alpha_{j+m}\beta\alpha_k \ldots \alpha_n\phi']$.

(5) if μ is none of the above, then if τ is the leftmost argument term occurring in ϕ, then where

(a) k is the number of free argument variables occurring before τ,

(b) ϕ' is the result of replacing the first occurrence of τ by a new variable β (with the same type as τ),

(c) $j = k + 1$, then
μ is the **jth-plugging of** $[\lambda\alpha_1 \ldots \alpha_{k-1}\beta\alpha_j \ldots \alpha_n\phi']$ **by** τ.

(6) If μ is none of the above, then

(a) ϕ is atomic

(b) $\alpha_1, \ldots, \alpha_n$ is the order in which these variables first occur in ϕ,

(c) $\mu = [\lambda\alpha_1 \ldots \alpha_n\rho^n\alpha_1 \ldots \alpha_n]$, for some term ρ^n, and

(d) μ is called **elementary**.

$\mathscr{I}$-assignments. If given an interpretation $\mathscr{I}$ of the language, an $\mathscr{I}$-assignment will be any function, f, defined on the primitive variables of the language such that when α is a variable of type t, $f(\alpha) \in \mathscr{D}_t$.

Denotations. If given an interpretation $\mathscr{I}$ and an $\mathscr{I}$-assignment f, we recursively define **the denotation of** term τ **with respect to** $\mathscr{I}$ **and** f("$d_{\mathscr{I},f}(\tau)$") as follows:

(1) where κ is any primitive name, $d_{\mathscr{I},f}(\kappa) = \mathscr{F}(\kappa)$

(2) where $\underline{\kappa}_\sigma$ is any closed sense term, $d_{\mathscr{I},f}(\kappa)=\mathscr{F}(\kappa)$

(3) where α is any primitive variable, $d_{\mathscr{I},f}(\alpha)=f(\alpha)$

(4) where $\underline{\kappa}^t_\sigma$ is any open sense term of type t,
$d_{\mathscr{I},f}(\underline{\kappa}^t_\sigma)=\mathit{sen}_{d_{\mathscr{I},f}(\sigma)}(\kappa^t)$

(5) where μ is an elementary λ-expression $[\lambda\alpha_1\ldots\alpha_n\rho\alpha_1\ldots\alpha_n]$
$d_{\mathscr{I},f}(\mu)=d_{\mathscr{I},f}(\rho)$

(6) where μ is the jth-plugging of ξ by τ,
$d_{\mathscr{I},f}(\mu)=\mathscr{PLUG}_j(d_{\mathscr{I},f}(\xi), d_{\mathscr{I},f}(\tau))$

(7) where μ is the jth-universalization of ξ, $d_{\mathscr{I},f}(\mu)=$
$\mathscr{UNIV}_j(d_{\mathscr{I},f}(\xi))$

(8) where μ is the j, k^{th}-conversion of ξ,
$d_{\mathscr{I},f}(\mu)=\mathscr{CONV}_{j,k}(d_{\mathscr{I},f}(\xi))$

(9) where μ is the j, k^{th}-reflection of ξ, $d_{\mathscr{I},f}(\mu)=\mathscr{REFL}_{j,k}(d_{\mathscr{I},f}(\xi))$

(10) where μ is the j, t'-vacuous expansion of ξ,
$d_{\mathscr{I},f}(\mu)=\mathscr{VAC}_{j,t'}(d_{\mathscr{I},f}(\xi))$

(11) where μ is the conditionalization of ξ and ζ,
$d_{\mathscr{I},f}(\mu)=\mathscr{COND}(d_{\mathscr{I},f}(\xi), d_{\mathscr{I},f}(\zeta))$

(12) where μ is the negation of ξ, $d_{\mathscr{I},f}(\mu)=\mathscr{NEG}(d_{\mathscr{I},f}(\xi))$

(13) where μ is the necessitation of ξ, $d_{\mathscr{I},f}(\mu)=\mathscr{NEC}(d_{\mathscr{I},f}(\xi))$

(14) where μ is any propositional formula ϕ, $d_{\mathscr{I},f}(\phi)$ is defined as follows:

(a) if ϕ is a primitive term of type p, $d_{\mathscr{I},f}(\phi)$ is already defined

(b) if $\phi=\rho^n\tau_1\ldots\tau_n$, $d_{\mathscr{I},f}(\phi)=$
$\mathscr{PLUG}_1(\mathscr{PLUG}_2(\ldots(\mathscr{PLUG}_n(d_{\mathscr{I},f}(\rho^n), d_{\mathscr{I},f}(\tau_n)),\ldots),$
$d_{\mathscr{I},f}(\tau_2)), d_{\mathscr{I},f}(\tau_1))$

(c) if $\phi=(\sim\psi)$, $d_{\mathscr{I},f}(\phi)=\mathscr{NEG}(d_{\mathscr{I},f}(\psi))$

(d) if $\phi=(\psi\rightarrow\chi)$, $d_{\mathscr{I},f}(\phi)=\mathscr{COND}(d_{\mathscr{I},f}(\psi), d_{\mathscr{I},f}(\chi))$

(e) if $\phi=(\forall\alpha^t)\psi$, $d_{\mathscr{I},f}(\phi)=\mathscr{UNIV}_1([\lambda\alpha^t\psi])$

(f) if $\phi=(\Box\psi)$, $d_{\mathscr{I},f}(\phi)=\mathscr{NEG}(d_{\mathscr{I},f}(\psi))$

(15) where *that*-ϕ is any complex propositional term, $d_{\mathscr{I},\mathscr{f}}(\textit{that-}\phi) = d_{\mathscr{I},\mathscr{f}}(\phi)$

(16) where $(\iota x^t)\phi$ is any object description, $d_{\mathscr{I},\mathscr{f}}((\iota x^t)\phi) =$

$$\begin{cases} o^t \text{ iff } (\exists \mathscr{f}')(\mathscr{f}'{}_{\dot{x}}\mathscr{f} \;\&\; \mathscr{f}'(x^t) = o^t \;\&\; \mathscr{f}' \text{ satisfies } \phi \text{ with respect to } w_0 \;\&\; (\mathscr{f}'')(\mathscr{f}''{}_{\dot{x}}\mathscr{f} \;\&\; \mathscr{f}'' \text{ satisfies } \phi \text{ with respect to } w_0 \rightarrow \mathscr{f}'' = \mathscr{f}')) \\ \text{undefined, otherwise} \end{cases}$$

(17) where $\underline{(\iota x^t)\phi}$ is any sense description, $d_{\mathscr{I},\mathscr{f}}(\underline{(\iota x^t)\phi}) = d_{\mathscr{I},\mathscr{f}}((\iota z^t)(F^{t/p})(zF \equiv F = [\lambda x\, \phi \;\&\; (y^t)(\phi^y_x \rightarrow y =_E x)]))$[10].

C. SATISFACTION

Given an interpretation $\mathscr{I}$ and an $\mathscr{I}$-assignment $\mathscr{f}$, we may define $\mathscr{f}$ **satisfies** ϕ **with respect to** w as follows:

(1) If ϕ is any primitive term of type p, $\mathscr{f}$ satisfies ϕ with respect to w iff $ext_w(d_{\mathscr{I},\mathscr{f}}(\phi)) = T$.

(2) If $\phi = \rho^n\tau_1 \ldots \tau_n$, $\mathscr{f}$ satisfies ϕ with respect to w iff $(\exists o_1)\ldots(\exists o_n)(\exists i^n)(o_1 = d_{\mathscr{I},\mathscr{f}}(\tau_1) \;\&\ldots\&\; o_n = d_{\mathscr{I},\mathscr{f}}(\tau_n) \;\&\; i^n = d_{\mathscr{I},\mathscr{f}}(\rho^n) \;\&\; \langle o_1, \ldots, o_n\rangle \in ext_w(i^n))$.

(3) If $\phi = \tau\rho$, $\mathscr{f}$ satisfies ϕ with respect to w iff $(\exists o)(\exists i)(o = d_{\mathscr{I},\mathscr{f}}(\tau) \;\&\; i = d_{\mathscr{I},\mathscr{f}}(\rho) \;\&\; o \in ext_{\mathscr{A}}(i))$.

(4) If $\phi = (\sim\psi)$, $\mathscr{f}$ satisfies ψ with respect to w iff $\mathscr{f}$ fails to satisfy ψ with respect to w.

(5) If $\phi = (\psi \rightarrow \chi)$, $\mathscr{f}$ satisfies ϕ with respect to w iff $\mathscr{f}$ fails to satisfy ψ with respect to w or $\mathscr{f}$ satisfies χ with respect to w.

(6) If $\phi = (\forall\alpha^t)\psi$, $\mathscr{f}$ satisfies ϕ with respect to w iff $(\forall \mathscr{f}')(\mathscr{f}'{}_{\dot{\alpha}}\mathscr{f} \rightarrow \mathscr{f}'$ satisfies ψ with respect to w).

(7) If $\phi = (\Box\psi)$, $\mathscr{f}$ satisfies ϕ with respect to w iff $(w')(\mathscr{f}$ satisfies ψ with respect to $w')$.

D. TRUTH UNDER AN INTERPRETATION

ϕ is **true under** $\mathscr{I}$ iff every $\mathscr{I}$-assignment $\mathscr{f}$ satisfies ϕ with respect to w_0.
ϕ is **false under** $\mathscr{I}$ iff no $\mathscr{I}$-assignment $\mathscr{f}$ satisfies ϕ with respect to w_0.

3. THE LOGIC

A. LOGICAL AXIOMS

In order to state some of the logical axioms, we will need to utilize the following three definitions, which are similar to their counterparts in the elementary and modal versions of the theory:

D_2 $\quad F^{t/p} = G^{t/p} =_{df} \Box(x^t)(xF \equiv xG)$

D_3 $\quad F^{(t_1,\ldots,t_n)/p} = G^{(t_1,\ldots,t_n)/p} =_{df} (n > 1)$
$(x^{t_2})\ldots(x^{t_n})([\lambda y^{t_1}\, Fyx^{t_2}\ldots x^{t_n}] = [\lambda y^{t_1}\, Gyx^{t_2}\ldots x^{t_n}])\ \&$
$(x^{t_1})(x^{t_3})\ldots(x^{t_n})([\lambda y^{t_2}\, Fx^{t_1}yx^{t_3}\ldots x^{t_n}] = [\lambda y^{t_2}\, Gx^{t_1}yx^{t_3}\ldots x^{t_n}])\ \&$
$\ldots \& (x^{t_1})\ldots(x^{t_{n-1}})([\lambda y^{t_n}\, Fx^{t_1}\ldots x^{t_{n-1}}y] = [\lambda y^{t_n}\, Gx^{t_1}\ldots x^{t_{n-1}}y]).$

D_4 $\quad F^p = G^p =_{df} [\lambda y^i\, F] = [\lambda y^i\, G].$

The logical axioms of our system are to be all of the modal closures of the following schemata, with the exception of the object description schemata, the unmodalized instances of which are to be axioms:

Propositional Schemata

LA1: $\phi \rightarrow (\psi \rightarrow \phi)$

LA2: $(\phi \rightarrow (\psi \rightarrow \chi)) \rightarrow ((\phi \rightarrow \psi) \rightarrow (\phi \rightarrow \chi))$

LA3: $(\sim\phi \rightarrow \sim\psi) \rightarrow ((\sim\phi \rightarrow \psi) \rightarrow \phi).$

Quantificational Schemata

LA4: (a) $(\alpha)\phi \rightarrow \phi^{\tau}_{\alpha}$, where τ contains no object descriptions and is substitutable for α[11]

(b) $(\alpha)\phi \rightarrow (\psi^{\tau}_{\beta} \rightarrow \phi^{\tau}_{\alpha})$, where ψ is any atomic formula, and τ both contains an object description and is substitutable for both α and β.

LA5: $(\alpha)(\phi \rightarrow \psi) \rightarrow (\phi \rightarrow (\alpha)\psi)$, provided α is not free in ϕ.

Modal Schemata

LA6: $\Box\phi \rightarrow \phi$

LA7: $\Box(\phi \rightarrow \psi) \rightarrow (\Box\phi \rightarrow \Box\psi)$

LA8: $\Diamond\phi \rightarrow \Box\Diamond\phi$

LA9: $\Box(\alpha)\phi \equiv (\alpha)\Box\phi$

LA10: $(x^t)(F^{t/p})(\Diamond xF \rightarrow \Box xF)$.

λ-Schemata

λ-EQUIVALENCE: where ϕ is any propositional formula with no object descriptions, the following is an axiom:

$$(x^{t_1})\ldots(x^{t_n})([\lambda\alpha_1\ldots\alpha_n\phi]x^{t_1}\ldots x^{t_n} \equiv \phi^{x^{t_1},\ldots,x^{t_n}}_{\alpha_1,\ldots,\alpha_n}).$$

λ-IDENTITY: where ρ is any relation term and $\alpha_1,\ldots,\alpha_n$, $\alpha'_1,\ldots,\alpha'_n$ are distinct object variables not free in ρ^p, the following is an axiom:

$$[\lambda\alpha_1\ldots\alpha_n\rho\alpha_1\ldots\alpha_n] = \rho \ \& \ [\lambda\alpha_1\ldots\alpha_n\rho^p] = [\lambda\alpha'_1\ldots\alpha'_n\rho^p].$$

Object Description Schemata

L-DESCRIPTIONS$_1$: where ψ is any atomic formula or conjunction of atomic formulas, the following is an axiom:

$$\psi^{(\iota x^t)\phi}_{\alpha} \rightarrow (\exists y^t)(\phi^y_x \ \& \ \psi^y_\alpha).$$

L-DESCRIPTIONS$_2$: where ψ is any atomic formula, the following is an axiom:

$$\psi^{(\iota x^t)\phi}_{\alpha} \rightarrow (y^t)(\phi^y_x \rightarrow \psi^y_\alpha).$$

L-DESCRIPTIONS$_3$: where ψ is any atomic formula with α_1 free and χ is any formula with α_2 free, the following is an axiom:

$$\psi^{(\iota x^t)\phi}_{\alpha_1} \rightarrow \sim((\exists y^t)(\phi^y_x \ \& \ \chi^y_{\alpha_2}) \ \& \ (\exists y^t)(\phi^y_x \ \& \ \sim\chi^y_{\alpha_2})).$$

In addition to the modal closures of the above schemata, and the unmodalized description schemata, the modal closures of the following five logical truths are also to be logical axioms:

LA11: $(F^{(t_1,\ldots,t_n)/p})(x^{t_1})\ldots(x^{t_n})(ExFx^{t_1}\ldots x^{t_n} \equiv Fx^{t_1}\ldots x^{t_n})$

LA12: $(F^p)(TrF^p \equiv F^p)$.

LA11 and LA12 are logically true because of our restrictions on the $\mathscr{F}$ function of interpretations (Section 2, A). They tell us that the explicit exemplification and truth predicates work as they should.

LA13: $\mathit{that}\text{-}\phi = \phi$

LA14: $A!^{t/p}\underline{\kappa}^t_\sigma \;\&\; R^{\langle t,t,i\rangle/p}\underline{\kappa}^t_\sigma \kappa\sigma$, where $\underline{\kappa}^t_\sigma$ is any sense name.

LA13 is logically true since in clause 15 of the definition of denotation$_{\mathscr{I},f}$, $\mathit{that}\text{-}\phi$ and ϕ are assigned the (semantically) same proposition. So $[\lambda y^i\ \mathit{that}\text{-}\phi]$ and $[\lambda y^i \phi]$ will be encoded by the same individuals.

Recall that we may read the second conjunct of LA14 as: the sense of the name "κ" with respect to σ represents κ to σ. LA14 is also logically true because of the restrictions placed on $\mathscr{F}$.

LA15: where ϕ is any propositional formula with no object descriptions, the following is an axiom:

$$(\iota x^t)\phi = (\iota z^t)(F^{t/p})(zF \equiv F = [\lambda x^t \phi \;\&\; (y^t)(\phi^y_x \rightarrow y =_{E^t} x)])^{12}.$$

LA15's validity is a consequence of clause 17 in the definition of denotation$_{\mathscr{I},f}$. Our sense descriptions denote Platonic Forms of type t (i.e., abstract objects of type t which encode a single t/p-property) which encode an individuating property (i.e., one which at most one object of type t can exemplify).

B. RULES OF INFERENCE

We use the same two rules of inference that were used in the earlier versions of the theory, $\rightarrow$E and UI. $\Box$I is still a derived rule, subject to the restriction discussed in Chapter IV, Section 3. We shall of course avail ourselves of the usual derived rules of inference and proof techniques.

RELATIONS is derivable in the same way it was derived in the earlier chapters and it is still subject to the restriction that the formulas ϕ used must not contain any object descriptions. Consequently, a type theory of relations falls right out of λ-EQUIVALENCE and $\boldsymbol{D}_3$. It is not a theory in which logically equivalent relations are identical. PROPOSITIONS is also derivable and subject to the same restriction as RELATIONS. PROPOSITIONS and $\boldsymbol{D}_4$ give us a theory of propositions.

We call the metaphysical system which consists of the interpreted typed language (without the unusual complex terms or distinguished predicates), together with LA1–LA10, the λ and object description schemata, and the rules of inference, the **typed object calculus**. The addition of the unusual complex terms and distinguished predicates, together with their semantics and logic (especially LA11–LA15), constitutes a special modification of the typed object calculus which has been designed specifically to deal with the data about propositional attitudes and mathematics.

4. THE PROPER AXIOMS

We assert that the modal closures of AXIOMS 1, 2, 3, 4, and 6 are all true A PRIORI, as well as the unmodalized instances of AXIOM 4. We insert a definition after the second axiom, in terms of which the third axiom is stated:

AXIOM 1. ("E^t-IDENTITY"): $x^t =_{E^t} y^t \equiv \Diamond E!^{t/p}x \,\&\, \Diamond E!^{t/p}y \,\&\, \Box(F^{t/p})(Fx \equiv Fy)$.

AXIOM 2. ("NO-CODER^t"): $\Diamond E!^{t/p}x^t \rightarrow \Box{\sim}(\exists F^{t/p})xF$.

Since we have general identity defined for all objects other than individuals, we need to say when two individuals are the same:

$$\boldsymbol{D}_5 \qquad x^i = y^i =_{df} x^i =_{E^i} y^i \vee (A!^{i/p}x \,\&\, A!^{i/p}y \,\&\, \Box(F^{i/p})(xF \equiv yF)).$$

The following axiom is therefore meant to govern any objects satisfying $\boldsymbol{D}_2$, $\boldsymbol{D}_3$, $\boldsymbol{D}_4$ or $\boldsymbol{D}_5$.

AXIOM 3. ("IDENTITY"): $\alpha = \beta \rightarrow (\phi(\alpha,\alpha) \equiv \phi(\alpha,\beta))$, where $\phi(\alpha,\beta)$ is the result of replacing some, but not necessarily all, free occurrences of α by β in $\phi(\alpha,\alpha)$, provided β is substitutable for α in the occurrences of α it replaces.

AXIOM 4. ("A-OBJECTS^t"): where ϕ is any formula in which x^t is not free, the following is an axiom:

$$(\exists x^t)(A!^{t/p}x \,\&\, (F^{t/p})(xF \equiv \phi)).$$

AXIOM 5. ("DESCRIPTIONS"): where ψ is any atomic formula with a free variable α of type t, the following is an axiom:

$$\psi_\alpha^{(\imath x^t)\phi} \equiv (\exists! y^t)\phi_x^y \,\&\, (\exists y^t)(\phi_x^y \,\&\, \psi_\alpha^y).$$

AXIOM 6. ("NECESSARY EXISTENCE"): for any type $t, t \neq i$, the following is an axiom:

$$\Diamond E!^{t/p}x^t \rightarrow \Box E!^{t/p}x^t.$$

E^t-IDENTITY, NO-CODER^t, IDENTITY, A-OBJECTS^t, and DESCRIPTIONS^t should be straightforward, given our familiarity with their counterparts in Chapters I–IV. Note that abstract objects of type t might encode abstract t/p-properties, as well as (possibly) existing ones.[13] The $F^{t/p}$-quantifier in A-OBJECTS^t ranges over all t/p-properties.

We have added one extra axiom to the typed theory to preserve the intuition that higher order objects are not contingent beings. Since higher order objects either possibly exist or fail to possibly exist, it follows from NECESSARY EXISTENCE that either they necessarily exist or they necessarily fail to exist. Philosophers who do not share the intuition that higher order objects are not contingent beings may not wish to embrace this axiom.

I think that E^t-IDENTITY, NO-CODER^t, IDENTITY, A-OBJECTS^t, DESCRIPTIONS^t, and NECESSARY EXISTENCE shall prove to be consistent.[14] We have taken steps to prevent the offending instances of property abstraction from being denoted. However, we have assumed that it is safe to have abstract objects of type t encode abstract properties of type t/p. I do not think that this move will introduce paradoxes, but it might. Should it do so, there are obvious ways to weaken the theory and preserve some of the applications which follow (we would, however, lose the very important model of mathematical entities). There is a great deal of investigation which must be carried out before we can feel confident that this particular version of the theory is consistent.

As usual, we add abbreviations for the appropriate English gerunds to our primitive vocabulary. And we add abbreviations for English proper names – names which are not necessarily associated with works of fiction.[15] Finally, we use the distinguished constants $B_1{}^{(i,p)/p}, B_2{}^{(i,p)/p}, \ldots$ to abbreviate the verbs of propositional attitudes such as believes, hopes, knows, expects, etc.[16]

CHAPTER VI

APPLICATIONS OF THE TYPED THEORY

1. MODELLING FREGE'S SENSES (I)

Frege's explanation, by way of ambiguity, of what appears to be the logically deviant behavior of terms in intermediate contexts is so theoretically satisfying that if we have not yet discovered or satisfactorily grasped the peculiar intermediate objects in question, then we should simply continue looking.

DAVID KAPLAN[1]

In this section, we translate and discuss the propositional attitude data which involve English names and definite descriptions that denote individuals. The data sentences are labelled (A)–(X), and because we are supposing with Frege that certain English terms occurring in them are ambiguous, there are several readings possible for each one. These readings are provided immediately after the particular datum is presented, and a discussion usually follows. In these discussions (in this section only), we revert to using the word "object" to refer to individuals and the word "property" to refer to properties of individuals (i.e., *i*/*p*-properties).

Also in these discussions, we shall modify somewhat the standard Fregean metalinguistic and metaphysical terminology. On the strict Fregean view, a term **expresses** its sense and **denotes** its denotation. And it is also said that the sense of a term **belongs to** the denotation of the term. Pictorially, these relationships are sometimes represented as follows:

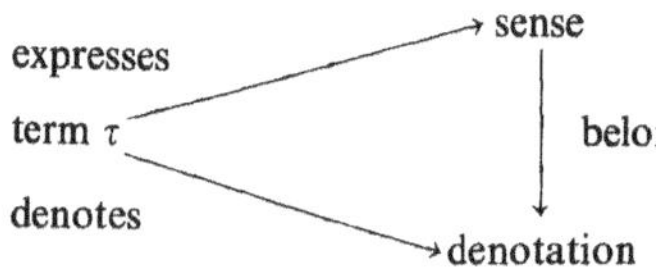

Now we shall talk as if terms do denote their denotations (this is made precise by our definition of $\text{denotation}_{\mathscr{I},f}$, V, Section 2), but we shall not suppose that terms "express" their senses. Instead, we shall talk about the *A*-object which is **associated with** the term **with respect to** an individual. Sometimes, we shall say that the *A*-object **serves as the sense of** the term **with respect to** a given individual. We assume with Frege that the sense of a term (especially proper names) varies from person to person. The special sense terms (and their interpretation) that we added to our language in Chapter V help us to represent this phenomenon and help to make the above terminology precise.

For reasons which will soon become apparent, we shall not talk in terms of the metaphysical "belonging to" relationship between senses and denotations. Instead, we shall talk about the weak correlates which the *A*-objects that serve as senses may have. Should the *A*-object have one or more weak correlates, we do not suppose that any of these objects necessarily serves as the denotation of the term in question. Diagrammatically, we get:

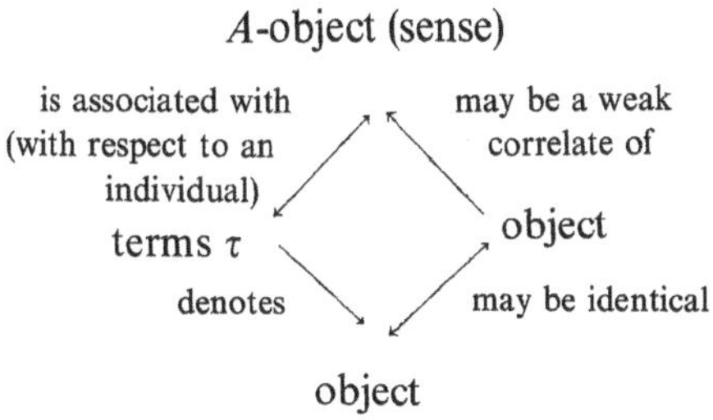

A. S believes that Lauben is late

(.1) *Bsthat-Ll* (DE RE)
(.2) $Bsthat\text{-}L\underline{l}_s$ (DE DICTO).

Suppose John feels ill one morning and resolves to stop in at the first physician's office he happens to pass on his way to work. He rounds a corner and sees a sign on a door:[2]

DR. GUSTAV LAUBEN

General Practioner

8:00 A.M. – 4:00 P.M.

At this point, he has now become part of a causal chain of events involving the name "Gustav Lauben".[3] He associates an *A*-object with this name – an *A*-object which serves as the sense of that name for him. We call that *A*-object "$\underline{\text{Lauben}}_{\text{John}}$". $\underline{\text{Lauben}}_{\text{John}}$ lends the name "Lauben" its cognitive significance or information value for John. It does so by encoding properties which serve to RE-present to him the object that he supposes is denoted by the name. $\underline{\text{Lauben}}_{\text{John}}$ may encode such properties as: being a (the) doctor whose office is at 15 High St., being a general practitioner, being the doctor whose signpost this is, being a doctor who works from eight to four, etc.

Some other person, *S*, who first encounters the name "Lauben" under different circumstances will not associate with this name the *A*-object John's utilized. The *A*-object *S* utilizes will encode properties presented to *S* as being characteristic of the object named "Lauben". Thus, $\underline{\text{Lauben}}_{S}$ lends the name "Lauben" a cognitive significance for *S* which is distinct from the cognitive significance this name has for John.

On the theory we have developed in Chapter V, it is axiomatic that $\underline{\text{Lauben}}_{S}$ represents Lauben for *S*. This is one way of capturing Frege's principle that the sense of a term account for its information value. Frege also required, however, that the sense of a term determine at most one object, and that this object, should there be one, serve as the denotation of the term. It requires additional semantic complexity to capture these Fregean principles, and were we interested in a more strict modelling of Frege's ideas, we could modify our semantics.[4] However, we've chosen not to place these constraints on senses because: (a) the successful explanation of the data on which we have chosen to work does not seem to require that we have such constraints on senses, and (b) there are cases which suggest that Frege's principles are too strong.

Suppose that the week before John went to Lauben's office, the Medical Review Board stripped Lauben of his license to practice, his medical school invalidated his degree, and he subsequently sold his office, never to return. It is just that no one bothered to take the sign down. In this situation, all of the properties which we have suggested might be encoded by $\underline{\text{Lauben}}_{\text{John}}$ are not exemplified by Lauben. Lauben is not the weak correlate of $\underline{\text{Lauben}}_{\text{John}}$; in fact, no object is.[5] Nevertheless, it seems reasonable to suppose that if John hasn't learned about Lauben's recent calamity, $\underline{\text{Lauben}}_{\text{John}}$, as described, still serves as the sense of "Lauben" and lends it cognitive significance.

To see this more clearly, suppose John knocks at the office door and no one answers. He notices that it is just after 8:00 A.M. He believes that Lauben is late (our datum sentence). His belief is not DE RE, since he believes this without believing that the friend of Leo Peters is late (suppose Lauben is Peter's unique friend). So his belief is DE DICTO.

Now even though Lauben is not the weak correlate of $\underline{\text{Lauben}}_{\text{John}}$, the latter *A*-object could still be instrumental in helping John to construct a proposition which serves as the "object" of his DE DICTO belief. We may suppose that the propositional object of his DE DICTO belief is $\mathscr{PLUG}_1(\mathit{being\ late}, \underline{\mathit{Lauben}_{John}})$. Had his belief been DE RE, the propositional object of his belief would have been $\mathscr{PLUG}_1(\mathit{being\ late}, \mathit{Lauben})$. His DE DICTO belief will be true just in case this latter proposition is true.[6] Since Lauben no longer comes to his office, his DE DICTO belief is a false one.

(In the above paragraph, we have switched typestyles and described the propositional objects of John's belief semantically. This allows us to graphically display the structure of these objects. In what follows, we frequently switch to the script typestyle when discussing either the propositional objects of beliefs or their constituents. With this warning, no confusion should arise.)

Consequently we seem to be able to describe the important facts of this situation without having to suppose that $\underline{\text{Lauben}}_{\text{John}}$ has a (unique) weak correlate, and without having to suppose that one of its weak correlates has to serve as the denotation of "Lauben". It is for this reason that we have chosen not to further complicate our semantics in order to present a more strict modelling of Frege's ideas.[7]

We have, however, validated another one of Frege's principles in the process – the *A*-object which serves as the sense of the term also serves as the denotation of the term inside DE DICTO belief contexts. To make this clearer, let us look at our two readings A.1 and A.2 in more detail. Our DE RE reading is $Bs\,that\text{-}Ll$ (A.1). From A.1, we can prove (1)–(4):

(1) $(\exists x)Bs\,that\text{-}Lx$

(2) $[\lambda x\, Bx\,that\text{-}Ll]s$

(3) $[\lambda x\, Bs\,that\text{-}Lx]l$

(4) $[\lambda xy\, Bx\,that\text{-}Ly]sl.$

On the assumption that Lauben exists, A.1 also implies (5):

(5) $(\exists y)(E!y\ \&\ Bs\,that\text{-}Ly).$

We have symbolized the DE DICTO reading of (A) as $Bs\,that\text{-}L\underline{l}_s$ (A.2). Thus, it is the sense of the name "Lauben" with respect to S which serves as the denotation of the name inside the DE DICTO belief context. From A.2, we may prove (6)–(10):

(6) $(\exists x)Bs\,that\text{-}Lx$

(7) $(\exists x)(A!x\ \&\ Bs\,that\text{-}Lx)$ (LA 14)

(8) $[\lambda x\, Bx\,that\text{-}L\underline{l}_s]s$

(9) $[\lambda x\, Bs\mathit{that}\text{-}Lx]\underline{L}_s$

(10) $[\lambda xy\, Bx\mathit{that}\text{-}Ly]s\underline{L}_s$

Quantification into the belief context works NORMALLY, as does λ-conversion.

An examination of another case should help – here is one inspired by Quine's work.[8]

B. Ralph believes that Cicero was a Roman

(.1) $Br\mathit{that}\text{-}Rc$ (DE RE)
(.2) $Br\mathit{that}\text{-}R\underline{c}_r$ (DE DICTO).

C. Ralph does not believe that Tully was a Roman

(.1) $\sim Br\mathit{that}\text{-}Rt$ (DE RE)
(.2) $\sim Br\mathit{that}\text{-}R\underline{t}_r$(DE DICTO).

D. Cicero is Tully

(.1) $c = t$
(.2) $c =_E t$.

The triad B-C-D (and other triads like it) constitutes a paradigm case where the English proper name exhibits logically deviant behavior. From B.1 and D.1 (or D.2), it follows that $Br\mathit{that}\text{-}Rt$. Identity elimination works normally. But from B.2 and D.1, nothing follows. And there is no reason to think that B.2, C.2, and (D) are jointly inconsistent. From B.2 and C.1, we can conclude both that $\underline{c}_r \neq t$ and that $R\underline{c}_r \neq Rt$. From B.2 and C.2, we can conclude both that $\underline{c}_r \neq \underline{t}_r$ and that $R\underline{c}_r \neq R\underline{t}_r$. Thus, we follow Frege in thinking that it is the ambiguity of the English proper name inside DE DICTO contexts which accounts for its logically deviant behavior.

We now precisely define the conditions under which someone has a true belief. Let us define **the erasure of** a formula ϕ ("ϕ^*") as the formula which results by deleting all the underlines and subscripts from terms occurring in ϕ. So where $\phi = R\underline{t}_r$, $\phi^* = Rt$. We now define:

***D*₆** *S truly believes that* ϕ ("$TBs\mathit{that}\text{-}\phi$") $=_{df} Bs\mathit{that}\text{-}\phi \;\&\; \phi^*$.

So from B.1 or B.2 and the supposition that Cicero was a Roman, it follows that Ralph has a true belief. From A.1 or A.2 and the supposition that Lauben is not late, it follows that S does not have a true belief.

Note that in the case of DE DICTO readings, S can truly believe that ϕ even when ϕ is false. In B.2, the propositional object of Ralph's belief is a false proposition – by our AUXILIARY HYPOTHESIS, no *A*-object exemplifies the (nuclear) property of being a Roman. This false proposition, however, is just a neutral object which helps Ralph to represent $\mathcal{PLUG}_1$ ($\mathcal{being\ a\ Roman}$, $\mathcal{Cicero}$). Consequently, we must abandon a certain principle some philosophers hold about true belief. The principle that *S* truly believes that ϕ iff *S* believes that ϕ and ϕ is true must be given up, not just because it is inconsistent with our treatment of belief, but also because doing so allows us to construe the logic of propositional attitude contexts as another application of the philosophical logic of encoding properties. The usefulness of abandoning the old principle is a good reason for doing so.

E. S believes that Lauben was mugged

(.1) *Bsthat-Ml* (RE)
(.2) $Bs\textit{that-}M\underline{l}_s$ (DICTO).

F. S′ believes that Lauben was mugged

(.1) *Bs′that-Ml* (RE)
(.2) $Bs'\textit{that-}M\underline{l}_{s'}$ (DICTO).

Suppose John goes to a party in the evening of the day he knocked on Lauben's door. Suppose also that Lauben is in good standing in the medical community, but that he just did not go to work that day. Leo Peters (Lauben is his unique friend, and John's unaware of their relationship) is there and John overhears him say "Dr. Lauben was mugged last night". The proposition John grasps when he hears this utterance is $\mathcal{PLUG}_1$($\mathcal{being\ mugged}$, $\underline{\mathcal{Lauben}}_{\mathcal{John}}$). His belief is DE DICTO, because he believes that Lauben was mugged without believing that the friend of Leo Peters was mugged. "$\underline{\mathcal{Lauben}}_{\mathcal{John}}$" may or may not be the (semantic) name of the *A*-object he associated with "Lauben" that morning. As John went through the day, he might have been involved in another context in which the name was used. The new information he gathers might get "encoded" by associating some distinct *A*-object which encodes all the old and new properties he now uses to re-present Lauben to him via the name "Lauben". For now, however, let us suppose that this name retains its earlier cognitive significance.

Now is there any reason to believe that the proposition John grasped when Peters uttered his sentence was the same proposition that Peters was entertaining? Suppose Peters' belief were DE DICTO. It seems like there would be "more intimate" properties encoded by $\underline{\text{Lauben}}_{\text{Peters}}$ than are encoded by $\underline{\text{Lauben}}_{\text{John}}$. We have supposed that Lauben is Peters' friend, and there might be a very complex *A*-object which Peters associates with "Lauben". Although the fact that $\underline{\text{Lauben}}_{\text{Peters}} \neq \underline{\text{Lauben}}_{\text{John}}$ is not a guarantee, it seems likely that the propositional object of Peters DE DICTO belief may differ from the object of John's DE DICTO belief.

Despite the fact these propositions may differ, there may still be good reason for thinking communication has taken place. A full discussion of how the communicative process operated in this situation would take us too far afield. We would have to discuss the intentions of the speaker to refer to Lauben, determine whether the speaker succeeded in referring to Lauben, and these might involve a discussion of the presuppositions of the context of the utterance. Even if we had a reasonable understanding of these features of the communicative process, it now seems in order to consider two further features. And they are, the degree to which $\underline{\text{Lauben}}_{\text{Peters}}$ and $\underline{\text{Lauben}}_{\text{John}}$ are "similar" *A*-objects and the kind of correspondence there is between the properties these *A*-objects encode and the properties Lauben exemplifies.

In the ideal case, $\underline{\text{Lauben}}_S$ and $\underline{\text{Lauben}}_{S'}$ will be identical (or one will be embedded in the other) and Lauben will be the unique weak correlate of both of them. At the other extreme, $\underline{\text{Lauben}}_S$ and $\underline{\text{Lauben}}_{S'}$ will have no properties in common and Lauben will be the unique weak correlate of neither of them. Communication takes place to a greater or lesser degree depending on whether the former or the latter of these two extremes is more closely approximated. So despite the fact that "Lauben was mugged" might be used by *S* to express one proposition and used by *S*′ to construct another proposition, communication between *S* and *S*′ takes place to a greater degree if both $\underline{\text{Lauben}}_S$ and $\underline{\text{Lauben}}_{S'}$ encode, for the most part, properties which Lauben exemplifies. In the cases where $\underline{\text{Lauben}}_S$ and $\underline{\text{Lauben}}_{S'}$ have little in common (with Lauben), communication is rather crude and not straightforward. Yet even in these latter cases, it is important to note that the language is holding everything together (as we might expect for DE DICTO beliefs). $\underline{\text{Lauben}}_S$ and $\underline{\text{Lauben}}_{S'}$ would have in common the fact that they are both associated with the name "Lauben". If Lauben was mugged, *S* and *S*′ have true beliefs.

G. S does not believe that the friend of Leo Peters was mugged

(.1) $\sim Bs\textit{that-}M(\iota x)Fxp$ (RE)

(.2) $\sim Bs\textit{that-}M\underline{(\iota x)Fxp}$ (DICTO).

Recall that we established that John's belief that Lauben was mugged was DE DICTO by the fact that he did not also believe that the friend of Leo Peters was mugged. But the English definite description exhibits logically deviant behavior inside belief contexts as well. On the DE RE reading of G, the proposition that he fails to believe is $\mathcal{PLUG}_1$ (*being mugged, the friend of Leo Peters*), i.e., $\mathcal{PLUG}_1$ (*being mugged, Lauben*). On the DE DICTO reading of (G), the proposition he fails to believe is $\mathcal{PLUG}_1$(*being mugged,* $\underline{\textit{the friend of Leo Peters}}$). The friend of Leo Peters is the abstract object which encodes just the property of being the friend of Leo Peters (by LA15). That is,

$$\underline{(\iota x)Fxp} = (\iota z)(G)(zG \equiv G = [\lambda x\, Fxp \;\&\; (y)(Fyp \rightarrow y =_E x)]).$$

The friend of Leo Peters serves as the sense of "the friend of Leo Peters". It lends the English description cognitive significance and information value. It also has at most one weak correlate, and in this case, its unique weak correlate happens to be the denotation of the description. Finally, the friend of Leo Peters serves as the denotation of the description when the description is inside a DE DICTO context.

H. Ralph believes that the man in the brown hat is a spy

(.1) $Brthat\text{-}S(\iota x)\phi_1$ (RE)
(.2) $Brthat\text{-}S\underline{(\iota x)\phi_1}$ (DICTO).

I. Ralph does not believe that the mayor of the town is a spy

(.1) $\sim Brthat\text{-}S(\iota x)\phi_2$ (RE)
(.2) $\sim Brthat\text{-}S\underline{(\iota x)\phi_2}$ (DICTO).

J. Ralph believes that the mayor of the town is not a spy

(.1) $Brthat\text{-}\sim S(\iota x)\phi_2$ (RE)
(.2) $Brthat\text{-}\sim S\underline{(\iota x)\phi_2}$ (DICTO).

K. Ortcutt is both the man in the brown hat and the mayor of the town

(.1) $o = (\iota x)\phi_1 \;\&\; o = (\iota x)\phi_2$
(.2) $o =_E (\iota x)\phi_1 \;\&\; o =_E (\iota x)\phi_2$.

If Ralph's belief in (H) is DE RE, the object of his belief is $\mathcal{PLUG}_1$(*being a spy, the man in the brown hat*), i.e., $\mathcal{PLUG}_1$(*being a spy, Ortcutt*). Given H.1 and (K), we may conclude (11):

(11) $Br\text{that-}So$.

Given I.1 and K, we may conclude (12):

(12) $\sim Br\text{that-}So$.

So H.1 and I.1 are inconsistent. Ortcutt himself is the constituent of the propositional object of the DE RE belief – the descriptions inside the relevant belief ascriptions "contribute" their denotation to the proposition. From H.1 and DESCRIPTIONS, we also get (13):

(13) $(\exists !y)(\phi_{1}{}^{y}_{x} \mathbin{\&} Br\text{that-}Sy)$.

However, let us suppose Ralph's belief is DE DICTO. The object of his belief is $\mathscr{PLUG}_1$(*being a spy*, *the man in the brown hat*). From H.2 and (K), nothing follows. From H.2 and I.2, it follows that the man in the brown hat ≠ the mayor of the town. If the mayor of the town is a spy, then it follows from H.2 that Ralph has a true belief. (14) also follows from H.2, given LA15 and DESCRIPTIONS:

(14) $(\exists !x)(A!x \mathbin{\&} (F)(xF \equiv F = [\lambda x\, \phi_1 \mathbin{\&} (y)(\phi_{1}{}^{y}_{x} \rightarrow y =_E x)])$
$\mathbin{\&} Br\text{that-}Sx)$.

Besides these, we have the usual consequences of H.2 based on existential introduction and λ-conversion:

(15) $(\exists x)Br\text{that-}Sx$

(16) $[\lambda x\, Bx\text{that-}S(\iota x)\phi_1]r$

(17) $[\lambda x\, Br\text{that-}Sx](\iota x)\phi_1$

(18) $[\lambda xy\, Bx\text{that-}Sy]r(\iota x)\phi_1$

Note that H.1 and J.1 ascribe contradictory beliefs to Ralph. Given (K), H.1 implies (11) and J.1 implies (19):

(19) $Br\text{that-}{\sim}So$.

From (11) and (19) we get (20):

(20) $(\exists F^0)(Br\text{that-}F^0 \mathbin{\&} Br\text{that-}{\sim}F^0)$.

However, (20) does not imply that Ralph believes a contradiction.

If H.2 and J.2 correctly describe Ralph's state of mind, then we cannot prove that Ralph has inconsistent beliefs. From H.2 and J.2, we cannot deduce that $(\iota x)\phi_1 \neq (\iota x)\phi_2$, but we can prove this from the plausible assumption that the property of being the man in the brown hat is distinct from the property of being the mayor of the town. Since $\underline{(\iota x)\phi}_1 \neq \underline{(\iota x)\phi}_2$, no substitutions into H.2 and J.2 would lead us to think that Ralph has inconsistent beliefs. However, from J.1 or J.2 and the fact that Ortcutt is a spy, we can prove that Ralph has a false belief, where,

S *falsely believes that* ϕ ("FB*sthat*-ϕ") $=_{df}$
B*sthat*-ϕ & $\sim TB$*sthat*-ϕ.

L. Ralph believes that the shortest spy is a spy

(.1) Br*that*-$S(\iota x)\phi_3$ (RE)
(.2) Br*that*-$S\underline{(\iota x)\phi}_3$ (DICTO)

If L.1 expresses what Ralph believes, then his belief would be of interest to the FBI. (21) follows from L.1:

(21) $[\lambda x\, Br$*that*-$Sx](\iota x)\phi_3$.

If Bond is the shortest spy, then (22) follows from L.1, and (23) follows from (21) or (22):

(22) Br*that*-Sb

(23) $[\lambda x\, Br$*that*-$Sx]b$.

If we assume that Bond exists and that an existence claim is built into ϕ_3, then we can generalize on (21)–(23) to get:

(24) $(\exists x)(E!x$ & Br*that*-$Sx)$

(25) $(\exists y)(E!y$ & $[\lambda x\, Br$*that*-$Sx]y)$.

All of this results because the propositional object of Ralph's belief has an existing object, namely Bond, as a constituent.

None of these results follow if L.2 expresses what Ralph believes. There is no way to use "exportation" on L.2 to produce (24) or (25). We can only reap the "standard" inferences from L.2 based on existential and lambda introduction:

(26) $(\exists x)Br$*that*-Sx

(27) $[\lambda x\, Br$*that*-$Sx]\underline{(\iota x)\phi}_3$,

(28) $(\exists y)(A!y$ & $[\lambda x\, Br$*that*-$Sx]y)$.

I take it that the FBI would not be interested by the fact that Ralph, like most everyone, uses an abstract object to represent whoever it is that is the shortest spy (in the absence of a DE RE belief).

M. Ralph believes someone is a spy

(.1) $(\exists x)Br\mathit{that}\text{-}Sx$

(.2) $Br\mathit{that}\text{-}(\exists x)Sx.$

M.1 and M.2 disambiguate (M). M.1 is similar to (24) and we might prefer to use the latter to read (M) properly. M.2 relates Ralph to a proposition which fails to have object constituents. No legitimate exportation on M.2 will get us to M.1.

N. S believes that Newton met Leibniz

(.1) $Bs\mathit{that}\text{-}Mnl$ (RE)

(.2) $Bs\mathit{that}\text{-}M\underline{n}_s l$ (DICTO/RE)

(.3) $Bs\mathit{that}\text{-}Mn\underline{l}_s$ (RE/DICTO)

(.4) $Bs\mathit{that}\text{-}M\underline{n}_s\underline{l}_s$ (DICTO).

In order to determine which of the readings of (N) is the correct one, we have to examine data triads to discover how the names "Newton" and "Leibniz" are functioning.

O. Frege believes that Hesperus is Hesperus.

P. Frege does not believe that Phosphorus is Hesperus.

Q. Phosphorus is Hesperus.

There are various ways to represent the triad O-P-Q. The preferred representation is as follows:

O.′ $Bf\mathit{that}\text{-}\underline{h}_f =_E h.$

P.′ $\sim Bf\mathit{that}\text{-}\underline{p}_f =_E h.$

Q.′ $p =_E h.$

Suppose Frege as a young man is being taught the names of the stars. Early one evening, his teacher points out Venus and says "That is Hesperus – it is the first visible star of the evening". Frege becomes, at that moment, part of an historical, causal chain of events connecting him with the name "Hesperus". So we suppose that Frege associates an abstract object sense with the name. $\underline{\text{Hesperus}}_{\text{Frege}}$ may encode: being the star to which my teacher is pointing, being the first visible star of evening, being clearly visible, being situated in position p in the western sky at 5:30 P.M. Thursday, December 7, 1860, etc. "Hesperus" would have a different cognitive value for someone who learned the name in different circumstances.

Now suppose Frege's teacher points out Venus to Frege early the next morning and says "That's Phosphorus – it is the last star visible in the morning". The young Frege will associate some new, distinct *A*-object with "Phosphorus". That's because the features of the learning situation are radically different. The object pointed out to him is in a position of the sky that appears unrelated to the position of the object pointed out the evening before. The names introduced are distinct. There is no reason for Frege to believe that the object pointed out to him then is identical with the object pointed out to him the evening before.

So if Frege's teacher does not tell him that Phosphorus is Hesperus, Frege could believe that Hesperus is Hesperus without believing that Phosphorus is Hesperus. Although there are various ways to represent this data as a consistent triad, we have chosen the reading on which Frege believes $\mathcal{PLUG}_1(\mathcal{PLUG}_2(\mathcal{I}\mathit{dentity}_E, \mathcal{H}\mathit{esperus}), \underline{\mathcal{H}\mathit{esperus}}_{\mathcal{F}\mathit{rege}})$ and fails to believe $\mathcal{PLUG}_1(\mathcal{PLUG}_2(\mathcal{I}\mathit{dentity}_E, \mathcal{H}\mathit{esperus}), \underline{\mathcal{P}\mathit{hosphorus}}_{\mathcal{F}\mathit{rege}})$.

R. John hopes that the strongest man in the world, whoever he is, beats up the man who just insulted him.

Preferred reading:

(R′) $\quad Hj\,\mathit{that}\text{-}B\underline{(\iota x)\phi_4}\,(\iota x)\phi_5$ (DICTO/RE).

On the preferred reading of (R), we interpret the first definite description as occupying a DE DICTO position and suppose that it contributes its sense to the proposition which is the object of John's hope.

S. Mary believes that the wife of Tully is the wife of Tully.

T. Mary doesn't believe that the wife of Cicero is the wife of Tully.

U. The wife of Cicero is the wife of Tully.

Preferred representation:[9]

S.′ $\quad Bm\,\mathit{that}\text{-}\underline{(\iota x)Wxt}_m =_E (\iota x)Wxt$ (DICTO).

T.′ $\sim Bm\mathit{that}\text{-}\underline{(\iota x)Wxc}_m =_E (\iota x)Wxt$ (DICTO).

U.′ $(\iota x)Wxc = (\iota x)Wxt.$

S-T-U is an interesting triad since it requires that we use the senses of the names "Tully" and "Cicero" with respect to Mary to construct the senses of the English descriptions "the wife of Tully" and "the wife of Cicero". That's because the wife of Tully and the wife of Cicero are identical, and so (S″) and (T″) are inconsistent.

S.″ $Bm\mathit{that}\text{-}\underline{(\iota x)Wxt} =_E (\iota x)Wxt$ (DICTO).

T.″ $\sim Bm\mathit{that}\text{-}\underline{(\iota x)Wxc} =_E (\iota x)Wxt.$

The wife of Tully and the wife of Cicero are identical because Cicero is Tully, and so being the wife of Tully just is being the wife of Cicero. Since these properties are identical, the object which encodes just being the wife of Tully is identical with the object which encodes just being the wife of Cicero. So we can not use (S″) and (T″) to help us understand how S-T-U is consistent because the proposition that the wife of Tully is identical$_E$ with the wife of Tully is identical with the proposition that the wife of Cicero is identical$_E$ with the wife of Tully.

So we MUST use the senses of "Tully" and "Cicero" with respect to Mary in order to suppose S-T-U is consistent. Thus, the wife of Tully$_m$ is a constituent of the propositional object of Mary's belief in (S′). Though the wife of Tully$_m$ could have at most one weak correlate, it fails to have any. By the AUXILIARY HYPOTHESIS, A-objects fail to exemplify the (nuclear) property of having a wife.[10]

Our definition of true belief still works fine:

(29) $$TBm\mathit{that}\text{-}\underline{(\iota x)Wxt}_m =_E (\iota x)Wxt \equiv$$

$$Bm\mathit{that}\text{-}\underline{(\iota x)Wxt}_m =_E (\iota x)Wxt \;\&\; (\iota x)Wxt =_E (\iota x)Wxt.$$

Given (S′) and given that there is a unique wife of Tully, it follows that Mary has a true belief.[11] If the negation of (T′) represented Mary's state of mind, she would still have a true belief. If Mary believes that the wife of Cicero was not the wife of Tully, and this was correctly represented as $Bm\mathit{that}\text{-}\underline{(\iota x)Wxc}_m \neq_E (\iota x)Wxt$, then she would have a false belief.

We insert here a few general remarks about our treatment of definite descriptions. G-L and R-U give us evidence for thinking that English descriptions have both a sense and a denotation. The sense of the definite description lends it cognitive value – a value to beings with representational capacities in that it enables them to recognize (or understand what it might be like to recognize) objects which have (never) been presented to them. If we are interested solely in describing the cognitive value of a given English description, we always have available to us a SENSE-description of our formal language. English descriptions do not come with their property terms marked as to whether the property denoted is exemplified or encoded by the object being described. And it might be that some other description of our language is better suited in having the intuitively right denotation of the English description. For example, we might prefer to use a description which contains an encoding subformula to translate "the student who killed an old moneylender," where this English description is meant to refer to Raskolnikov of *Crime and Punishment* (Chapter IV, Section 4). But insofar as we are interested purely in the phenomenon of the apparent deviant behaviour of this English description inside DE DICTO contexts, it might just be that the student who killed an old moneylender serves well enough as its cognitive value. This depends on whether there is conclusive data which shows that the question of getting the denotation right and the question of explaining apparent deviant behavior are not independent.[12]

V. Bill believes that there is a barber who shaves all and only those who do not shave themselves.

W. Bill does not believe that the sun is shining and the sun is not shining.

X. Necessarily, there is a barber who shaves all and only those who do not shave themselves iff the sun is shining and the sun is not shining.

(V′) $Bb\mathit{that}\text{-}(\exists x)(Bx \;\&\; (y)(Sxy \equiv {\sim}Syy))$

(W′) ${\sim}Bb\mathit{that}\text{-}(Ss \;\&\; {\sim}Ss)$

(X′) $\Box((\exists x)(Bx \;\&\; (y)(Sxy \equiv {\sim}Syy)) \equiv (Ss \;\&\; {\sim}Ss))$.

Were we to identity propositions with their extensions (i.e. with functions from possible worlds to truth values), all contradictions and necessary falsehoods would be identified. That's because contradictions and necessary falsehoods are false at all possible worlds, and there is a unique function which maps all the worlds to the truth value: False. This model of propositions has the unfortunate consequence that if we believe a proposition *P* we believe all propositions *Q* which are necessary equivalent to P. In particular, if we believe some necessary falsehood then we believe all contradictions, since they are the same "proposition". But surely, V-W-X count as data and give us good evidence for thinking that necessarily equivalent propositions may be distinct. These considerations provide overriding reasons for choosing the style of semantics we have employed. In fact, it follows from (V′) and (W′) that the propositions in question are distinct, since (V′) and (W′) imply (30):

(30) $(\exists x)(Bx \,\&\, (y)(Sxy \equiv \sim Syy)) \neq (Ss \,\&\, \sim Ss)$.

2. MODELLING FREGE'S SENSES (II)

We now consider the propositional attitude data triads which involve English terms that denote higher order objects.

A. John believes that Woodie is a woodchuck

(.1) $Bjthat\text{-}Ww$ (DE RE)
(.2) $Bjthat\text{-}\underline{W}_j w$ (DE DICTO)[13].

B. John does not believe that Woodie is a groundhog

(.1) $\sim Bjthat\text{-}Gw$ (DE RE)
(.2) $\sim Bjthat\text{-}\underline{G}_j w$ (DE DICTO).

C. Being a woodchuck just is being a groundhog

(.1) $G = W$

A.1 and B.1 are inconsistent, given C.1. A.1 asserts that John believes the proposition $\mathscr{PLUG}_1$(*being a woodchuck*, *Woodie*), whereas B.1 asserts that John does not believe $\mathscr{PLUG}_1$(*being a groundhog*, *Woodie*). But since (C) is a true identity statement which asserts that the properties of being a woodchuck and being a groundhog are identical, it's provable that these propositions are identical. So either A.1 or B.1 must be false.

A.2 and B.2 can both be true together, however. A.2 asserts that John believes $\mathscr{PLUG}_1$($\underline{\textit{being a woodchuck}}_{\mathit{John}}$, *Woodie*). $\underline{\text{Being a woodchuck}}_{\text{John}}$ is the abstract property of individuals (*i*/*p*-property) which serves as the sense of the name "being a woodchuck" with respect to John. The $\mathit{sen}^{i/p}_{\mathit{John}}$ function of our semantics (Chapter V, Section 2, A) assigns "being a woodchuck" a member of $\mathscr{A}_{i/p}$ (that is, a member of the abstract objects of type i/p). So "being a woodchuck$_{\text{John}}$" denotes $\mathit{sen}^{i/p}_{\mathcal{John}}$("being a woodchuck").

Being a woodchuck$_{\text{John}}$ can be plugged up with any individual – our $\mathcal{PLUG}$ function is defined so that it operates on all i/p-properties. Consequently, $\mathcal{PLUG}_1(\underline{\mathit{being\ a\ woodchuck}}_{\mathcal{John}}, \mathcal{W}\mathit{oodie})$ is a type p object and can serve as the propositional object of someone's belief.

B.2 asserts that John does not believe $\mathcal{PLUG}_1(\underline{\mathit{being\ a\ groundhog}}_{\mathcal{John}}, \mathcal{W}\mathit{oodie})$. Being a groundhog$_{\text{John}}$ is the abstract i/p-property which serves as the sense of "being a groundhog" with respect to John. If A.2 and B.2 are true, it follows both that $\mathcal{PLUG}_1(\underline{\mathit{being\ a\ woodchuck}}_{\mathcal{John}}, \mathcal{W}\mathit{oodie}) \neq \mathcal{PLUG}_1(\underline{\mathit{being\ a\ groundhog}}_{\mathcal{John}}, \mathcal{W}\mathit{oodie})$ and that $\underline{\mathit{being\ a\ woodchuck}}_{\mathcal{John}} \neq \underline{\mathit{being\ a\ groundhog}}_{\mathcal{John}}$. So as Frege predicted, the senses of the property terms flanking the identity sign in C are distinct.

This seems right – "woodchuck" and "groundhog" probably entered John's vocabulary under different circumstances. Maybe on one occasion he saw and was told he was seeing a woodchuck. He then utilized an A-object which encoded properties he took to be characteristic of the property of being a woodchuck. And maybe on another occasion, someone described woodchucks to him improperly, in the process saying only that he was describing an animal called a "groundhog". John would not have known that in fact these properties are the same. Sentence (C) would be informative to him.

D. John believes that the chair in front of the class is Crayola crayon blue

(.1) $Bjthat\text{-}CCB(\imath x^i)\phi_1$ (RE)
(.2) $Bjthat\text{-}\underline{CCB}_j(\imath x^i)\phi_1$ (DICTO)

E. John does not believe that the chair in front of the class is French fire engine blue

(.1) $\sim Bjthat\text{-}FEB(\imath x^i)\phi$ (RE)
(.2) $\sim Bjthat\text{-}\underline{FEB}_j(\imath x^i)\phi_1$ (DICTO)

F. Crayola crayon blue just is French fire engine blue

(.1) $FEB = CCB$

D-E-F is analyzed analogously with A-B-C. We may suppose 'French fire engine blue" and "Crayola crayon blue" to be names of the same shade of blue. So (F) is an informative identity statement about properties. As a boy, John may have become directly acquainted with this property. But the label on his Crayola crayons just read "blue", and he has never seen that shade of blue labeled "French fire engine".

G. John believes that Bill has the property of being a student

(.1) $Bjthat\text{-}HasSb$ (RE)
(.2) $Bjthat\text{-}\underline{Has}_jSb$ (DICTO).

H. John does not believe that Bill exemplifies the property of being a student

(.1) $\sim Bjthat\text{-}ExSb$ (RE)
(.2) $\sim Bjthat\text{-}\underline{Ex}_jSb$ (DICTO).

I. Having a property just is exemplifying a property

(.1) $Ex = Has$

G-H-I might describe a student beginning in philosophy, unaware of the technical sense philosophers have for the word "exemplifies". "Has" and "exemplifies" both denote relations of type $(i/p, i)/p$. We suppose that there are abstract objects of this type which serve as the senses of these names with respect to John. G.2 asserts that John believes $\mathcal{PLUG}_1$ $(\mathcal{PLUG}_2$ $(\underline{\mathit{having}}_{\mathit{John}}, \mathit{Bill}), \mathit{being\ a\ student})$; whereas H.2 asserts that John doesn't believe $\mathcal{PLUG}_1$ $(\mathcal{PLUG}_2$ $(\underline{\mathit{exemplifying}}_{\mathit{John}}, \mathit{Bill}), \mathit{being\ a\ student})$.

To handle our next triad, J-K-L, we need to add some functional notation to our language:

> where ρ is a term of type $(t_1, t_2)/p$ and τ is a term of type t_1, then $\rho(\tau)$ is a term of type t_2.

Let us interpret this notation as follows

$$d_{\mathcal{I},f}(\rho(\tau)) = d_{\mathcal{I},f}((\iota\alpha^{t_2})\rho\tau\alpha).$$

So $\rho(\tau)$ is the object which is such that τ bears ρ to it. Using this interpreted notation, we might construe adverbs as names of relations of type $(i/p, i/p)/p$, i.e., relations which relate two i/p-properties. For example, "slowly" might denote a relation between the property of walking and the property of walking slowly. In the language, "slowly" combines with "walk" to form "slowly (walk)", which denotes the property of walking slowly.

This gives us a means of representing J-K-L consistently.

J. John believes Bill walked bravely....

(.1) Bj*that*-$B(W)b$ (RE)
(.2) Bj*that*-$\underline{B}_j(W)b$ (DICTO).

K. John does not believe that Bill walked courageously....

(.1) $\sim Bj$*that*-$C(W)b$ (RE)
(.2) $\sim Bj$*that*-$\underline{C}_j(W)b$ (DICTO).

L. Walking bravely...just is walking courageously...

(.1) $B(W) = C(W)$

Examples like G-L should demonstrate that our analysis for DE DICTO belief is generalizable throughout the types. This treatment of beliefs about higher order objects suggests a solution to the "paradox" of analysis. Central to this puzzle are data triads similar to the ones we've been discussing. Here is an example:

M. It is trivial that the concept brother is identical with the concept brother.

N. It is not trivial that the concept male sibling is identical with the concept brother.

O. The concept brother is identical with the concept male sibling.

Although there are various ways to state the puzzle precisely, all we need to say is that the puzzle involves the question of how an identity statement like (O) can be (an) informative (analysis). Philosophers who believe that property terms denote sets and express properties, and who hold that "brother" and "male sibling" express the same property are left with no means of accounting for the informative nature of the identity statement formed by flanking an identity sign with the property DENOTING terms "the concept brother" and "the concept male sibling". What is to serve as the senses of these expressions?

We suppose here that property analyses are sentences which say that two apparently distinct properties are identical. We simply extend Frege's view of their informative character by supposing that there are distinct abstract properties which serve as the senses of the above property denoting expressions. In order to represent M-N-O correctly, we need to note that strictly speaking, triviality is person relative – what is trivial for one person may not be trivial for another. We assume that triviality is a relation between persons and propositions. Consequently, we introduce "T" to be a name of type $(i, p)/p$ and we read "Tx*that*-ϕ" as: it is trivial for x that ϕ. This forges an analogy with the other propositional attitudes. Terms which follow the "it is trivial for x" prefix behave like terms in propositional attitude contexts – sometimes they denote their senses.

Clearly, if we're restricting overselves to a discussion of a particular individual S, then (P′)-(Q′)-(R′) would be the proper way to capture the triad P-Q-R:

P. It is trivial for S that the concept brother is identical with the concept brother.

Q. It is not trivial for S that the concept male sibling is identical with the concept male sibling.

R. The concept brother is identical with the concept male sibling.

P.′ $\quad Ts\mathit{that}\text{-}\underline{B}_s =_E B$

Q.′ $\quad \sim Ts\mathit{that}\text{-}\underline{MS}_s =_E B$

R.′ $\quad MS = B.$

The proposition asserted to be trivial by (P) is $\mathscr{PLUG}_1(\mathscr{PLUG}_2(\mathit{identity}_E, \mathit{being\ a\ brother}), \underline{\mathit{being\ a\ brother}}_s)$. Its triviality derives from the logical truth that $\underline{\text{being a brother}}_s$ represents being a brother to S. The proposition that's not trivial according to (Q′) is $\mathscr{PLUG}_1(\mathscr{PLUG}_2(\mathit{identity}_E, \mathit{being\ a\ brother}), \underline{\mathit{being\ a\ male\ sibling}}_s)$.

(P′)-(Q′)-(R′) may be a good account of P-Q-R, but the original triad was M-N-O. How are we to represent it? Well, since the English prefix "it is trivial" as it occurs in (M) is not relativized to a particular individual, it seems that (M) asserts that it is trivial for everyone that the concept brother is identical with the concept brother. The relevant reading of (N) seems to be: everyone is such that it is not trivial for them that the concept male sibling is identical with the concept brother. If we recall that we have allowed primitive variables of type i to serve as subscripts for sense terms, then (M′)-(N′)-(O′) seems to be the correct way to translate the data triad:

M.′ $\quad (x)Tx\mathit{that}\text{-}\underline{B}_x =_E B$

N.′ $\quad (x) \sim Tx\mathit{that}\text{-}\underline{MS}_x =_E B$

Q.′ $\quad MS = B.$

An even closer representation of M-N-O would be one which uses the iota-operator to capture the English definite article. Let us represent "the concept brother" as $(\iota y^{i/p})(y = B)$ and "the concept male sibling" as $(\iota y^{i/p})(y = MS)$. A consistent reading of M-N-O would be:

M.″ $(x)Tx\mathit{that}\text{-}(\iota y^{i/p})(y = \underline{B}_x) =_E (\iota y^{i/p})(y = B)$.

N.″ $(x){\sim}Tx\mathit{that}\text{-}(\iota y^{i/p})(y = \underline{MS}_x) =_E (\iota y^{i/p})(y = B)$

O.″ $(\iota y^{i/p})(y = MS) = (\iota y^{i/p})(y = B)$.

3. MODELLING IMPOSSIBLE AND FICTIONAL RELATIONS

A. IMPOSSIBLE RELATIONS

Sometimes we think about "impossible" individuals. These are not individuals which are such that some contradiction is true. Rather, these are individuals like the$_{\mathscr{A}}$ round square, which encode incompatible properties. We can also think about "impossible" relations – the symmetrical non-symmetrical relation is one. Here is an A PRIORI truth about this relation:

(1) The symmetrical non-symmetrical relation is symmetrical.

We cannot analyze "the symmetrical non-symmetrical relation" as a description involving exemplification formulas since it would fail to denote. There are no higher order objects which exemplify both being symmetrical and being non-symmetrical. (1) would be turned into a falsehood.

However, we may read the description as "the object which encodes symmetricality and non-symmetricality", and then generate the A PRIORI truth that this object encodes symmetricality. Let us suppose that the properties in question are of type $((i, i)/p)/p$.[14] We then have the following instance of A-OBJECTS:

$$(\exists x^{(i,i)/p})(A!^{((i,i)/p)/p}x \;\&\; (F^{((i,i)/p)/p})(xF \equiv F = S \vee F = \bar{S}))$$

This axiom, plus the definition of identity among $(i, i)/p$-objects, justifies our talking about THE abstract relation which encodes being symmetrical and being non-symmetrical. As in Chapter II, we define:

the$_{\mathscr{A}}$ symmetrical non-symmetrical relation $=_{df}$
$(\iota x^{(i,i)/p})(A!x \;\&\; (F)(xF \equiv F = S \vee F = \bar{S}))$,

dropping the obvious typescripts. It is then provable that the$_{\mathscr{A}}$ symmetrical non-symmetrical relation encodes symmetricality. This theorem represents (1).

The analysis proposed for data about the$_{\mathcal{A}}$ round square seems, therefore, to generalize within type theory to the data about the$_{\mathcal{A}}$ symmetrical non-symmetrical relation.

B. FICTIONAL RELATIONS

When we see a sentence like "Einstein discovered that there is no such thing as simultaneity", how are we to understand it? Did Einstein discover that no two events ever exemplify the relation of simultaneity? Or did he discover that the simultaneity relation does not exist?[15] I'm not sure how to decide the issue, but the latter reading seems a legitimate option. We might therefore suppose that simultaneity is a fictional relation, and for our present purposes, we could suppose that it is native to the Newtonian (science) fiction. So let us identify it in a way analogous to our earlier work.

Let us suppose that events are special kinds of propositions, and that they are type p objects. A relation among two events would be of type $(p,p)/p$. Newtonian mechanics presupposes that simultaneity is a (possibly) existing relation of type $(p,p)/p$. However, in our view, it must be an abstract relation. Let "s" be a name of type $(p,p)/p$ which denotes this relation. We analyze (2) as (2)′:

(2) Einstein discovered that simultaneity does not exist

(2)′ $De_{that\text{-}\sim E!!s}$[16]

But which abstract relation does "s" denote? Well, by typing the definitions of character ("*Char* (x^t,s)") and native ("*Native* (x^t,s)"), we may construct a typed *N*-CHARACTERS axiom such as:[17]

$$(x^t)(s)(Native\,(x,s) \rightarrow x = (\iota z^t)(F^{t/p})(zF \equiv \Sigma_s(Fx)).$$

Let "n" denote the story of Newtonian mechanics. We then get the following instance of *N*-CHARACTERS:

$$Native\,(s,n) \rightarrow s = (\iota z)(zF \equiv \Sigma_n Fs),$$

where "s" denotes simultaneity and is not a restricted variable. This may prove to be an interesting way to identify other fictional relations of disproven scientific theories.

Could someone write a story about non-scientific, fictional relations? Could we dream about non-existent relations? If these are genuine possibilities, we will have further data for the application of our type theory.

4. MODELLING MATHEMATICAL MYTHS AND ENTITIES

In this final section of applications, we conjecture somewhat about the farthest reaches of metaphysical hyperspace. Thus far, we have found applications for abstract individuals which encode possibly existing properties, and abstract properties (relations) which encode properties of properties (relations). If we now recall that the theory even asserts that there are abstract individuals which encode abstract properties, then it is only natural to wonder whether we can find applications for these recondite creatures. It also seems natural to suggest that mathematical entities just are such abstract individuals. We devote the rest of this section to spelling out this suggestion.

The first thing to do is to get clear on the data. We look for true sentences of natural language whose truth seems to require that there be mathematical objects to serve as the denotations of certain terms found in them. The following sentences must surely count as such:

(1) In Peano number theory, zero is not a successor of any number.

(2) In Peano number theory, there is a prime number greater than two.

(3) John wondered whether the number one hundred seventeen thousand, four hundred and sixty seven is prime.

(4) In Zermelo-Fraenkel set theory, there is a set which has no members.

(5) In Zermelo-Fraenkel set theory, for any given property and set, there is a second set whose members are just those members of the given set exemplifying the given property.

(6) In Zermelo-Fraenkel set theory, every transitive set has a transitive power set.

A few comments are in order. The locutions "in Peano number theory" and "in Zermelo-Fraenkel set theory" are NOT meant to be short for "in the standard model of Peano number theory" or "in the standard model of ZF". Rather, we pretheoretically understand these locutions as a kind of "in the story" prefix and analogize the situation to that of fiction. We will try to make this pretheoretical understanding of these locutions precise by supposing that mathematicians author mathematical stories. The stories are usually communicated to other mathematicians by declaring a few basic principles of the stories (the axioms of the particular mathematical theory) with the understanding that all necessary consequences of the principles are also to be part of the story. If the story is rich enough, then it is of interest to other mathematicians to try to discover other truths according to the story. We are supposing that it is a contingent fact that there are any mathematical stories or storytellers.

Failure to appreciate the fact that sentences like (1)–(6) are the basic data has resulted in a confused debate about the ontological status of mathematical objects. The "platonists" who accept Tarski's theory of truth both mistakenly suppose mathematical assertions unprefixed by the story operator, like "there is a prime number greater than two," are literally true, and also fail to distinguish the quantifier "there is" from "there exists". Consequently, they conclude, for example, from the fact that the sentence "there is a prime number greater then two" is literally true, that there exists a (mathematical or metaphysical) object which satisfies the open sentence "x is a number and x is prime and x is greater than two". This conclusion seems unwarranted. On the other hand, the nominalists who accept Tarski's theory of truth simply deny that such unprefixed sentences are ever true and proceed to try to show that we need not make an essential appeal to the truth of such sentences in any of our subsequent (scientific) theorizing or problem solving. But such a strategy simply doesn't account for all the data. For example, we get true sentences like (1), (2), (4), (5), and (6) by prefixing the story operator to the theorems of mathematics. And, of course, there are data like (3) (one cannot discard data like (3) as being part of the problem of propositional attitude contexts). Any serious attempt to account for the truth of the above data on a compositional basis must attribute semantic significance to the number words in these sentences in such a way that these words have the same significance when they occur in sentences like (2) as they do when they occur in sentences like (3).

Now that we've identified the data, we next identify the denotations of all the names appearing in the data by utilizing and extending the machinery developed in IV., Section 4 and VI., Section 3. Let us concentrate initially on datum sentence (1).

First, we identify the story in question. Let us use "PNT" to abbreviate "Peano's Number Theory". Given STORIES (IV., Section 4), and the supposition that PNT is a story, we may assert:

$$PNT = (\iota z)(F)(zF \equiv (\exists F^0)(\Sigma_{PNT} F^0 \;\&\; F = [\lambda y\, F^0])).$$

That is, Peano's number theory is that abstract individual which encodes just the vacuous properties constructed out of propositions true according to Peano's number theory. We therefore know, in principle, what the denotation of "Peano's number theory" is as it occurs in (1). We have a clearer idea of which propositions are true according to mathematical stories than we do in the case of literature. For Peano's number theory, we know that the conjunction of (7)–(11) is true according to the story:

(7) Zero is a number.

(8) Every number has a successor which is a number.

(9) No two distinct numbers have the same successor.

(10) Zero is not the successor of any number.

(11) If zero has some property, and a number's having that property implies that its successor has that property, then every number has that property.

In addition, a very strong principle governs the mathematical story operator; where "*s*" ranges over mathematical stories:[18]

$$\text{MATH-}\Sigma\text{-SUB}: (F^0)(G^0)(\Sigma_s F^0 \ \& \ F^0 \Rightarrow G^0 \rightarrow \Sigma_s G^0).$$

So, all the necessary consequences of the conjunction of (7)–(11) are also true according to Peano's number theory.

What about the denotations of the other names in (1)? There are three names to consider: zero ("0"), number ("*N*"), and successor ("*S*"). If we assume that particular numbers are individuals which may exemplify properties but are not themselves exemplified by anything, then "zero" names an individual. Which individual? Well, for one thing, it is a native character of the story. Given *N*-CHARACTERS, we may conclude:

$$0 = (\iota z)(F)(zF \equiv \Sigma_{PNT} F0).$$

So zero encodes the following sorts of properties (which are properties it exemplifies according to the story): being a number, not being the successor of any number, being less than all other numbers, etc.

So far, this is a straightforward application of the theory we elaborated earlier. But the interesting new twist comes as we try to identify the denotations of "number" and "successor". We regard the property of being a number as a native character of the mathematical story as well. Consequently, it is an abstract, rather than existing, property. It is not the kind of property that could exist. Which abstract property? Consider the following instance of typed N-CHARACTERS:

$$N = (\imath z^{i/p})(F^{(i/p)/p})(zF \equiv \Sigma_{PNT} FN).$$

That is, being a number is that abstract property which encodes just the properties it exemplifies according to the story. So it encodes the following sorts of properties: being a property that the number zero exemplifies ($[\lambda F^{i/p} F0]$), being a property such that everything that exemplifies it has a successor which also exemplifies it ($[\lambda F^{i/p}(x)(Fx \rightarrow (\exists y)(Fy \ \& \ Syx))]$), being a property such that zero is not the successor of anything which exemplifies it ($[\lambda F^{i/p}(x)(Fx \rightarrow \sim S0x)]$).

The successor relation is also an abstract relation which is a character of the story:

$$S = (\imath x^{(i,i)/p})(F^{((i,i)/p)/p})(zF \equiv \Sigma_{PNT} FS).$$

Consequently, this relation encodes: the property of being a relation which zero fails to bear to any other number ($[\lambda F^{(i,i)/p} \sim (\exists x)(Nx \ \& \ F0x)]$), the property of being a relation such that for every number, there is a second number which bears it to the first ($[\lambda F^{(i,i)/p}(x)(Nx \rightarrow (\exists y)(Ny \ \& \ Fyx))]$), etc.

We have now identified, in principle, the denotations of the names occurring in (1), and (7)–(11). We may translate these sentences into our formal language, knowing what the significance of each term is, as follows:

(1)′ $\Sigma_{PNT} \sim (\exists x)(Nx \ \& \ S0x)$

(7)′ $N0$

(8)′ $(x)(Nx \rightarrow (\exists y)(Ny \ \& \ Syx))$

(9)′ $(x)(y)(Nx \ \& \ Ny \ \& \ x \neq y \rightarrow \sim (\exists z)(Nz \ \& \ Szx \ \& \ Szy))$

(10)′ $\sim (\exists x)(Nx \ \& \ S0x)$

(11)′ $(F)(F0 \ \& \ (x)(y)(Nx \ \& \ Ny \ \& \ Syx \ \& \ Fx \rightarrow Fy) \rightarrow (z)Fz)$.

In datum sentence (2), we have three terms, "even", "prime", and "greater than", which are definable using the primitives of the theory. So we could regard (2) as an abbreviation of a truth-of-Peano-number-theory in which all the defined terms have been eliminated. But it may be preferable to regard terms like "even", "prime", and "greater than", as denoting complex abstract relations. There is an interesting field of investigation here – the semantics is quite prepared for accommodating complex relations which are constructed (using the logical functions) out of simpler abstract relations. Also, the abstraction schema for relations remains neutral on the question of whether the relations constructed or the constructing relations must be possibly existing or abstract relations. So there seems to be no reason why we can't suppose "E", "P", and "$>$" in the following translation of (2) denote complex abstract properties and relations:

(2)′ $\Sigma_{PNT}(\exists x)(Nx \mathbin{\&} Ex \mathbin{\&} Px \mathbin{\&} x > 2)$.

Datum sentence (3) is now translatable, given this understanding of (2). One may plug abstract individuals into abstract (complex) relations to produce propositions. Propositions may serve as the objects of belief. It may just be that the object of John's wonder in (3) is the proposition which results by plugging the abstract individual 117, 467 into the abstract relation of being prime. So we translate (3) as follows:[19]

(3)′ $Wj\textit{whether-}P117,467$.

Next, we consider (4) and (5). Given our discussion of Peano's number theory, the identification of the denotations of the terms occurring in them should be straightforward. Zermelo-Fraenkel set theory is a mathematical story – it encodes just the vacuous properties constructed out of propositions necessarily implied by the conjunction of the axioms of the theory (as they are formulated in standard second order predicate logic). The property of being a set ("S") is an abstract property which is a character of the mathematical story. The membership relation ("$\in$") is an abstract relation which is also a character of the story. These characters may be identified in a manner analogous to the identification of the abstract property of being a number and the abstract successor relation. And, (4) and (5) may be translated as follows:

(4)′ $\Sigma_{ZF}(\exists x)(Sx \mathbin{\&} (y)y \notin x)$

(5)′ $\Sigma_{ZF}(F)(x)(Sx \rightarrow (\exists y)(Sy \mathbin{\&} (z)(z \in y \equiv z \in x \mathbin{\&} Fz)))$.

(4)′ and (5)′ are both true since they say that the NULLSET and SUBSET axioms are true according to ZF. Sentences like these guarantee that there are lots of abstract individuals which exemplify-according-to-ZF the property of being a set. It also follows that (4)′ and (5)′ are necessarily true.

Finally, consider (6). The property of being transitive is a property of sets just as being prime is a property of numbers. It is therefore a complex, abstract property. It is constructed, using the logical functions, out of two other abstract relations, the membership relation and the subset relation (which is also complex). Consider the following two λ-expressions:

(a) $[\lambda x^i y^i \ (z^i)(z \in x \rightarrow z \in y)]$

(b) $[\lambda x^i \ (y^i)(y \in x \rightarrow y \subseteq x)]$

(a) yields a complex relation. When this such relation holds between objects x and y, we write "$x \subseteq y$". Of course, no two objects bear this relation to one another, but lots of objects bear this relation to one another according to ZF. We know further that since this relation is a native character of ZF, that it is abstract and encodes just the properties it exemplifies according to ZF.

(b) yields the complex property of being transitive ("T"). Again, no objects exemplify this property, but many objects exemplify it according to ZF. And since it too is a native character of the story, it may be identified as that abstract property which encodes just the properties that the property of being transitive exemplifies according to ZF. Now if we let "P" abbreviate $[\lambda xy \ (u)(u \in x \equiv u \subseteq y)]$, we may then translate (6) as:

(6)′ $\Sigma_{ZF}(x)(Tx \rightarrow (\exists y)(Pyx \ \& \ Ty))$.

The truth conditions for (6)′ are compositional and should be precise.

By translating just these six pieces of data, we should have given the reader a good idea of how we would answer questions that have faced philosophers in the other, traditional approaches to this topic. We've presented a view on the nature of mathematical truth, for example, simply by identifying the data as we have. If the translations are successful, not much more need be said on this question, since the metaphysical theory provides us with lots of consequences of the translations. These consequences will be unacceptable only to those who already have some favorite philosophy of mathematics. Another question which has turned out to be quite a conundrum arises in connection with model theory. Given that there are several ways to construct models of number theory in set theory, which of the structures that do the modelling really are the numbers? We close with a brief discussion on this topic.

The question in fact seems a little misguided. All that the model-theoretic facts show is that mathematicians could have carried on without a separate story about the numbers. But as long as the story of numbers can be considered to be distinct from the story of sets, numbers and sets have separate ontological status. The stories just have different characters. Model-theoretic facts about the possible models of a theory have no bearing on the question of what the mathematical entities ARE. This is a question decided by a metaphysical theory. Indeed, no model-theoretic facts about the interpretation of our metaphysical theory change any of the definitions or theorems telling us about the nature of possible worlds, monads, or fictional characters. Such facts only tell us whether the theory is consistent, whether the logic is strong enough to deduce the consequences of the theory, or whether the theory forces a categorical structure for its interpretation, etc.

We have attempted to produce an explanation of what mathematical entities and stories are entirely within the framework of metaphysics. We have done so by forging a strong link between myth and mathematics. Our account helps us to explain the semantic significance of number words (in so far as they are used as nouns or names) as they occur in natural language, as well as the semantic significance of other special words of mathematics. And we have verified what mathematicians have claimed all along, namely, that sets, numbers, etc., are abstract objects. And this should come as no surprise to those who take metaphysics to be ontologically prior to mathematics.

There is one last puzzling group of data: non-mathematical statements of number which take the form "there are n F's." For example, "there are two planets." A pure logicians' analysis of this example is: there are distinct objects x and y which are both planets and all other planets are identical with x or y. We know such a procedure can be generalized. But an important alternative view is: number words in non-mathematical statements of number denote "natural" numbers. And there is a "natural" way to identify such numbers using the metaphysical and logical machinery of our system, without appealing to any mathematical notions: let "zero" denote the A-individual which encodes just the properties which fail to be exemplified; let "one" denote the A-individual which encodes just the properties which are uniquely exemplified; etc. Then, a metaphysical analysis of "there are nine planets" would be: nine encodes the property of being a planet. This "DE RE" reading is necessarily true if true, as opposed to the pure logicians' contingent "DE DICTO" reading, which is compatible with the truth that it's possible that there are ten planets. The metaphysical analysis could be an example of a necessary, A POSTERIORI truth. Both readings appear to be legitimate options and reveal a subtle ambiguity in the data. I commend these considerations to my readers.

CONCLUSION

I do not plan to specifically argue that I have established the thesis of the book, namely, that our research program is a progressive one. It is for the reader to decide, on the basis of the material presented in Chapters II, IV, and VI, whether the theory anchoring the program both helps us to explain our data and, together with the auxiliary hypotheses, predicts hitherto unknown, novel facts. Nor do I intend to argue here that the theory generates more interesting consequences and can correctly represent more interesting pieces of knowledge than any of its competitors. Readers with a firm grasp on (the outstanding data facing) other research programs must decide which provides the most natural, elegant, and unified treatment of the variety of basic problems tackled here. Instead, I would like to end with a few paragraphs in which I conjecture about the even greater variety of future research possibilities which flower in the foregoing formal garden. I will identify issues which may be of interest to logicians, pure and applied metaphysicians, linguists, and pure and applied epistemologists.

We may have provided logicians with a new "primary interpretation" for second order logic. Since we have a precise, axiomatic theory of relations, there is no longer any reason to suppose that the n-place relation names and variables of second order languages denote (range over) sets. Interpretations in which they denote relations seem to me metaphysically prior and preferable to the traditional primary interpretations. Consequently, standard incompleteness results for the second order predicate calculus, which are based on interpretations in which the domain of n-place relations just is the power set of the nth Cartesian product of the domain of objects, do not apply to the elementary object calculus (or to any second order predicate calculus interpreted in a similar manner). We must look to the theory of relations to give us the basic facts about the (size of the) realm of relations. This approach to the semantics of second order languages should provide a wide range of metatheoretic research possibilities (I have tried to formulate the model theoretic semantics as elegantly as possible to facilitate future work on metatheoretical questions).

For those who prefer pure metaphysics to model theory, the most promising new line of research concerns time. There should be a way to model instants of time in much the same way that we have modelled possible worlds. If the basic tense operators are added to our language (and interpreted in the normal way as quantifiers over an ordered set of times, taken as primitive), the definition of a world should convert into a definition of a time. That is because the diamond operator in the second clause of the definition takes on new significance. Semantically, this clause would tell us "...and at some world and time, z encodes just the vacuous properties constructed out of propositions true there and then". All of the theorems about worlds should also convert into theorems about times. The interesting project would be to then try to reconstruct the worlds by finding the relevant A-objects which harness all the instants of time that occur at a given world.

There are other research possibilities for the pure metaphysician, since many traditional and modern issues can be reanalyzed from our new perspective. For example, traditional ontological arguments could be reanalyzed in a way which does justice to their logic, since the theory of A-objects makes it possible to coherently reason about a thing without prejudice as to whether that thing exists. Arguments may fail because the exemplification-encodes ambiguity infects premises involving singular predications. The theory however would not provide support for any A PRIORI argument which concludes that THERE EXISTS an object which exemplifies every property God exemplifies according to the story.

Having a precisely demarcated background ontology might also make worthwhile a reexamination of such modern questions as the ontological status of kinds, minds, works of art, etc. Could a case be made for thinking that these types of entities are species of abstract objects, or must they be assigned separate ontological status? Pure metaphysicians may ask about our theory of identity. Is it general enough? Also, I have tried to make suggestions along the way for extending the theory which may be worth pursuing. The theory may be rich enough for others to create new definitions, discover interesting consequences, add new auxiliary hypotheses, etc.

In addition to traditional questions about reference, meaning, synonymy, analyticity, and realism (all about which a cogent view may naturally evolve from the theory), applied metaphysicians (for example, philosophers of language) might be interested in the systems which result by adding context dependent names to our syntax. Context dependent names would receive both a denotation and a sense relative to each context of utterance. It might be possible to render pronouns such as "I", "he", "she", "you", and "it" as such context dependent names. This would help us explain why we can't eliminate such pronouns in certain belief contexts through substitution of co-referential proper names, since it might be the senses of such expressions (relative to the given context) which are involved rather than their denotations.

Also, the development of a Montague style fragment of English would be of interest to philosophers of language and linguists. Since the data we considered consisted of true English sentences, it becomes important to show that our formal languages have enough resources to provide a general semantic treatment of all such sentences (I have relied on a certain tradition of translation in the applications). The resulting system would differ from Montague's in many important respects. Provisions would have to be made for the structural ambiguity we may have discovered in singular predication. English names and descriptions would be ambiguous in belief contexts, given our Fregean solution to the puzzles of belief. Belief contexts would not involve the same scope ambiguities which Montague relies on to differentiate DE RE and DE DICTO beliefs, though certain other scope ambiguities will still be present. Since monads mirror their worlds and consistent stories may be systematically related to worlds, it might prove theoretically useful to dispense with existing objects in the immediate interpretation of English. That is, it might simplify the rules of interpretation. The idea is that proper names of existing individuals could denote their monads and names and (in)definite descriptions of discourse objects could denote characters of stories. At present, however, this may just be a wild conjecture and a great deal of research must be carried out before we could determine whether such a project is going to pan out. However, I do not think that we'll encounter the same problem of finite representability which faces Montague's systems, since the terms of our language denote entities which are NOT functions defined on infinite domains.

This brings us to the last area for future investigation we will discuss – epistemology. Since thinking of, dreaming about, searching for, and worshipping objects all may involve a complex relationship to *A*-objects, pure epistemologists may consider postulating a basic kind of acquaintance relationship which might serve to ground these intentional relations. The idea is that we analyze worshipping Zeus, searching for the fountain of youth, thinking about Hamlet, etc., in terms of acquaintance with these objects plus different intellectual (possibly propositional) attitudes we adopt toward them. This relationship of acquaintance is not a causal one, though we may come to bear this relation to A-objects, in part, through causal interactions with copies of novels, storytellers, etc. (The causal theory of names seems to me to fail. A-objects do not have spatial location, and so no "dubbing", in the customary sense, of an A-object ever takes place. But I hope we have provided a good enough case for thinking that some proper names denote A-objects).

Nor is this acquaintance relationship the same kind of intellectual acquaintance we have with properties. We must certainly be acquainted with properties like being red, being round, being sharp, etc. We may all agree that no matter what world we are placed in, we could recognize whether or not an object in that world was red, round, sharp, etc. Yet supposing that we are acquainted with properties seems to me to be a natural explanation of this fact. These distinct acquaintance relationships may serve as the basis for appending an epistemology to our metaphysics. I think that a reasonable epistemology can be found, as long as we do not suppose that we can gather knowledge about abstract objects in so far as they are objects. The only proper knowledge that we can have with respect to abstract objects is A PRIORI. But this does not rule out the fact that we can gather knowledge about them in so far as they are fictional characters, for example.

Applied epistemologists who investigate procedural models of semantic competence and performance might avail themselves of abstract objects as well. In building a program which models the acquisition of proper names, maybe we should insert a subroutine which collates the available assertions involving the name being introduced and associates with the name the A-object which encodes just the properties denoted by the predicates of such assertions. Such A-objects might serve as constituents of propositions to be constructed when processing future assertions involving the name. I am not sure how the mind works, but in storytelling, the assimilation of each new assertion involving the name or co-referential pronoun may involve a switch to a new abstract object which codes up the properties jointly involved in the new and previous assertions. A full procedural representation of discourse might therefore involve entire sequences of abstract objects.

Of course, what is needed now is a group assault. It is my hope that these suggestions are worthy of such a group enterprise.

APPENDIX A

MODELLING THE THEORY ITSELF

This appendix will be divided into three parts. In part A, we discuss the paradoxes of encoding and their joint solution. In part B, we describe an extensional model of the monadic portion of the elementary theory, suggested by Dana Scott. In part C, we describe Scott's model of the monadic portion of the modal theory.

PART A

Just as in set theory, unrestricted abstraction schemata lead to paradox. However, in the theory of abstract objects, it is the joint operation of two unrestricted schemata which proves to be inconsistent, *A*-OBJECTS and λ-EQUIVALENCE. λ-EQUIVALENCE has been restricted indirectly, and RELATIONS has been restricted directly, so as to avoid these paradoxes. There are two paradoxes to consider, one by Romance Clark and the other by Alan McMichael, and they both stem from a common source. It seems to me important to sketch the proofs so that the reader may see how they arise from the unrestricted versions of our axioms.

Suppose we dropped the two major restrictions on λ-formation and RELATIONS (i.e., the restrictions imposed by the definition of propositional formula). We could then form the following two λ-expressions: $[\lambda x(\exists F)(xF \,\&\, {\sim}Fx)]$ ("encoding a property that is not exemplified"), and $[\lambda x(F)(xF \rightarrow Fx)]$ ("exemplifying every property that's encoded"). Alternatively, we could produce instances of RELATIONS as follows:

$$(\exists F)(x)(Fx \equiv (\exists G)(xG \,\&\, {\sim}Gx))$$

$$(\exists F)(x)(Fx \equiv (G)(xG \rightarrow Gx)).$$

But then consider the following argument ("Clark's paradox"), first reported in Rapaport [1976], p. 225:

Consider the abstract object a_0 which encodes just $[\lambda x(\exists F)(xF \,\&\, {\sim}Fx)]$, and suppose it exemplifies $[\lambda x(F)(xF \rightarrow Fx)]$. By λE, it follows that $(F)(a_0F \rightarrow Fa_0)$, so a_0 must exemplify $[\lambda x(\exists F)(xF \& \sim Fx)]$ as well as encode it. Again, by λE, $(\exists F)(a_0F \& \sim Fa_0)$, i.e., $\sim(F)(a_0F \rightarrow Fa_0)$. But then a_0 must fail to exemplify $[\lambda x(F)(xF \rightarrow Fx)]$ (by λ-EQUIVALENCE and $\equiv$E), contrary to hypothesis.

So suppose a_0 fails to exemplify $[\lambda x(F)(xF \rightarrow Fx)]$. Then $\sim(F)(a_0F \rightarrow Fa_0)$, i.e. $(\exists F)(a_0F \& \sim Fa_0)$. Call this property "$R$" and note also that by λI, a_0 exemplifies $[\lambda x(\exists F)(xF \& \sim Fx)]$. Since a_0 encodes just one property, R must be $[\lambda x(\exists F)(xF \& \sim Fx)]$. But by definition of R, a_0 fails to exemplify R, i.e., $\sim[\lambda x(\exists F)(xF \& \sim Fx)]a_0$, contradiction. ⊠

A second contradiction would also be provable because we could form the following λ-expression: $[\lambda y\, y=x]$ ("being identical to x"), where this abbreviates a much longer λ-expression with encoding subformulas and relation quantifiers. Again, by RELATIONS, we would know that there is such a property. But then consider the following argument ("McMichael's paradox"), first reported in a footnote to our [1979b]:

By A-OBJECTS, we have that $(\exists x)(A!x \& (F)(xF \equiv (\exists u)(F = [\lambda y\, y=u] \& \sim uF)))$. Call this object a_1 and consider the property $[\lambda y\, y=a_1]$. Assume that a_1 encodes $[\lambda y\, y=a_1]$. By definition of a_1, we know $(\exists u)([\lambda y\, y=a_1]=[\lambda y\, y=u] \& \sim u[\lambda y\, y=a_1])$. Call this object a_2. So, $[\lambda y\, y=a_1]=[\lambda y\, y=a_2] \& \sim a_2[\lambda y\, y=a_1]$. By $=$I, we know $a_1=a_1$, and by λI, we know $[\lambda y\, y=a_1]a_1$. Since $[\lambda y\, y=a_1]=[\lambda y\, y=a_2]$, it follows by $=$E that $[\lambda y\, y=a_2]a_1$. So by λE, $a_1=a_2$. But then, $\sim a_1[\lambda y\, y=a_1]$ (from the definition of a_2 and $=$E), contrary to hypothesis.

So suppose that $\sim a_1[\lambda y\, y=a_1]$. By definition of a_1, $\sim(\exists u)([\lambda y\, y=a_1]=[\lambda y\, y=u] \& \sim u[\lambda y\, y=a_1])$. That is, $(u)([\lambda y\, y=a_1]=[\lambda y\, y=u] \rightarrow u[\lambda y\, y=a_1])$. But since $[\lambda y\, y=a_1]=[\lambda y\, y=a_1]$, it follows that $a_1[\lambda y\, y=a_1]$. Contradiction. ⊠

It is doubtful that the source of these paradoxes lies with the presence of relation quantifiers in λ-expressions. Logicians have not found any special trouble with the second order predicate calculus, in which one finds relations defined with quantification over other relations. For example, here's a standard instance of the relations schema of the second order predicate calculus:

$$(\exists F)(x)(Fx \equiv (\forall G)Gx).$$

This asserts that there is a property of "having all properties". This property would be denoted using "$[\lambda x(\forall G)Gx]$". Properties such as these do not seem to cause any special consistency problems. The *only* reason for adding the "no relation quantifiers" restriction on λ-expressions is that given the style of semantics we have employed, it is rather complicated to interpret such expressions without the resources of type theory (in type theory, we suppose that "$[\lambda x(\forall G)Gx]$" abbreviates "$[\lambda x(\forall G)ExGx]$", where "$Ex$" is a predicate which denotes the exemplification relation between a property and an object which exemplifies it. We then interpret this latter λ-expression as $\mathcal{UNIV}_1(d_{\mathcal{I},f}(Ex))$ i.e., as the first universalization of this exemplification relation).

Consequently, the elimination of encoding subformulas from the abstraction schema for relations seems to be the most theoretically satisfying way of avoiding the paradoxes. McMichael first suggested this move to me while I was writing [1979a], though at the time it turned out to be insufficient. That was because the language which was being used, had the logical notion of identity as a primitive. When McMichael discovered his paradox while we were writing [1979b], we realized that we would have to place extra restrictions on λ-EQUIVALENCE and RELATIONS. We had to banish primitive identity formulas from these schemata as well as encoding formulas. However, early in 1980, I discovered that identity for properties and relations could be plausibly defined in terms of encoding formulas. The key to this discovery was $\boldsymbol{D}_3$ (Chapter I, the definition for relation identity), which was forged during a search for a complete theory of relations to accompany the theory of abstract objects. As long as one uses the defined notion of relation identity and eliminates the primitive logical notion from the language, one cannot generate a relation of identity. $[\lambda xy\, x = y]$ is ill-formed and the definiens in $\boldsymbol{D}_3$ cannot be used as the formula ϕ in RELATIONS. So it turns out that the paradoxes stem from a common source, since the elimination of encoding formulas alone from the abstraction schema for relations suffices to prevent both paradoxes.

PART B

Dana Scott has suggested the following extensional model of the monadic portion of the elementary theory. By an "extensional" model, we mean one in which properties and relations with the same exemplification extensions are identical. Although the theory does not require this identification, doing so facilitates model construction a great deal without calling the consistency results into question. However, as a project for future research, we should look for models of the theory in which such an identification is not made.

Since we are simplifying matters by considering only the monadic portion of the theory, we shall construct a model in which the following two axioms are true:

> $(\exists F)(x)(Fx \equiv \phi_o)$, where F is not free in ϕ_o and ϕ_o has no encoding formulas
> $(\exists x)(F)(xF \equiv \phi)$, where x is not free in ϕ.

We need not concern ourselves with the existence predicate or with NO-CODER since in the following interpretation, only "abstract" objects will do any encoding.

The interpretation is constructed in ZF plus individuals. Let us call our set of individuals "$\mathscr{E}$", so that it corresponds with terminology defined at the end of Section 2, Chapter I. We will use "e" as variables ranging over the members of $\mathscr{E}$. Now let "$\mathscr{R}$" be the set which is the union of two copies of the power set of $\mathscr{E}$, where the members of the first copy are coded with a plus (or a one) and the members of the second copy are coded with a minus (or a zero). That is,

$$\mathscr{R} = \{+\} \times \mathfrak{P}(\mathscr{E}) \cup \{-\} \times \mathfrak{P}(\mathscr{E}).$$

The members of $\mathscr{R}$ will be our properties, and we use "r" as variables ranging over them. Also, we use "s" as variables ranging over the members of the power set of $\mathscr{E}$. So, for every r, there is a unique s such that either r is identical with $\langle +, s \rangle$ or r is identical with $\langle -, s \rangle$. Also let us say that the absolute value of r is just the set s such that $r = \langle +, s \rangle$ or $r = \langle -, s \rangle$. We write "$|r|$" to designate the absolute value of r. Finally, let "$\mathscr{A}$" be the power set of $\mathscr{R}$. These will be our abstract objects, and we use "a" as variables ranging over these objects.

Now in terms of these sets, we can specify the interpretation which we suspect is a model and say what exemplification and encoding amount to. Recall that an interpretation of the theory will be the form $\langle \mathscr{D}, \mathscr{R}, ext_{\mathscr{R}}, ext_{\mathscr{A}}, \mathscr{F} \rangle$. So let $\mathscr{D}$ be the union of $\mathscr{E}$ and $\mathscr{A}$, and as usual, we use "o" to range over the members of $\mathscr{D}$. Let $\mathscr{R}$ be just the $\mathscr{R}$ defined immediately above. And in terms of this picture, we now state the conditions under which an object exemplifies and/or encodes a property:

(1) $o \in ext_{\mathscr{R}}(r)$ iff either
  (i) $(\exists e)(o = e \;\&\; (\exists s)((r = \langle +, s \rangle \vee r = \langle -, s \rangle) \;\&\; e \in s))$, or
  (ii) $(\exists a)(o = a \;\&\; (\exists s)(r = \langle +, s \rangle))$

(2) $o \in ext_{\mathscr{A}}(r)$ iff either
  (i) $(\exists e)(o = e \;\&\; r \neq r)$, or
  (ii) $(\exists a)(o = a \;\&\; r \in a)$.

So clause (1i) ensures that an existing object exemplifies a property $\mathscr{r}$ just in case it is an element of $|\mathscr{r}|$. Clause (1ii) ensures that an abstract object exemplifies a property $\mathscr{r}$ iff $\mathscr{r}$ is a plus-marked set of $\mathscr{e}$'s. Since this condition is vacuous with respect to the abstract objects, either all abstract objects exemplify a given property or none do. Clause (2i) ensures that no existing objects encode properties, since the condition $\mathscr{r} \neq \mathscr{r}$ is never satisfied. Clause (2ii) ensures that an abstract object $\mathscr{a}$ encodes a property $\mathscr{r}$ just in case $\mathscr{r}$ is an element of $\mathscr{a}$.

Now we need to prove the following claim: that both $(\exists F)(x)(Fx \equiv \phi_o)$ and $(\exists x)(F)(xF \equiv \phi)$ are true under this interpretation. The claim can be demonstrated quite easily once the following fact and lemma are seen to be true. The fact is that "$(x)\phi$" is true iff both $(\mathscr{e})\phi'$ and $(\mathscr{a})\phi'$ are both true, where ϕ' is the semantic translation of ϕ. Clearly, universal claims of the object language are true just in case they hold with respect to all existing objects and with respect to all abstract objects. The lemma we need can be stated as follows:

Invariance Lemma: Given the above interpretation, let an assignment to the variables, $\mathscr{f}$, be fixed. Then, where ϕ_o is an any formula with no encoding formulas where x is free,

$$(\exists \mathscr{f}')(\exists \mathscr{a})(\mathscr{f}'_{\bar{x}}\,\mathscr{f} \;\&\; \mathscr{f}'(x) = \mathscr{a} \;\&\; \mathscr{f}' \text{ satisfies } \phi_o) \equiv$$
$$(\mathscr{f}')(\mathscr{a})(\mathscr{f}'_{\bar{x}}\,\mathscr{f} \;\&\; \mathscr{f}'(x) = \mathscr{a} \rightarrow \mathscr{f}' \text{ satisfies } \phi_o).$$

If we allow ourselves to talk derivatively about objects satisfying formulas (instead of strictly talking in terms of assignments to the variables satisfying formulas), then our lemma may be read quite simply: an abstract object satisfies a formula ϕ_o with no encoding formulas iff all abstract objects do. This is a consequence of clause (1ii) above, which places an vacuous condition on an abstract objects' exemplifying a property. Since the details of the proof are somewhat messy, we save the proof of this lemma until the end. For now, let's suppose that it's true.

Then consider the first axiom: $(\exists F)(x)(Fx \equiv \phi_o)$, where ϕ_o has no encoding formulas and no free F's. Semantically, we have to show: $(\exists \mathscr{r})((\mathscr{e})(\mathscr{e} \in \mathscr{ext}_{\mathscr{R}}(\mathscr{r}) \equiv \phi'_o) \;\&\; (\mathscr{a})(\mathscr{a} \in \mathscr{ext}_{\mathscr{R}}(\mathscr{r}) \equiv \phi'_o))$, in view of the above fact (again, ϕ'_o is the semantic version of ϕ_o). So let's describe a way to choose $\mathscr{r}$ so that the conjunction of universal claims is true. Take an arbitrary ϕ'_o and let $\mathscr{s}$ be the set of existing objects satisfying ϕ'_o. Clearly there must be such an $\mathscr{s}$, since we have all the possible subsets of $\mathscr{E}$ to choose from (we are appealing here to the fact the following instance of the SUBSET axiom of ZF governs our interpretation: $(\exists y)(z)(z \in y \equiv z \in \mathcal{E} \ \& \ \phi_o'))$. Now take an arbitrary abstract object, say a_5. If a_5 satisfies ϕ_o', then choose i to be $\langle +, s \rangle$. Clearly, if an arbitrary e is an element of $ext_{\mathcal{R}}(i)$, then $e \in s$, by the above clauses for exemplification. So e satisfies ϕ_o'. And if e satisfies ϕ_o', then e is an element of s, and so $e \in ext_{\mathcal{R}}(i)$. So, $(e)(e \in ext_{\mathcal{R}}(i) \equiv \phi_o')$. Since a_5 satisfies ϕ_o', all abstract objects do. And since i is $\langle +, s \rangle, (a)(a \in ext_{\mathcal{R}}(i))$. So $(a)(a \in ext_{\mathcal{R}}(i) \equiv \phi_o')$. Consequently, our conjunction of universal claims is true.

On the other hand, if a_5 fails to satisfy ϕ_o', choose i to be $\langle -, s \rangle$. Again, $(e)(e \in ext_{\mathcal{R}}(i) \equiv \phi_o')$, since if an arbitrary $e \in ext_{\mathcal{R}}(i)$, then $e \in s$ and e satisfies ϕ_o' (and vice versa). Since a_5 fails to satisfy ϕ_o', no abstract object satisfies ϕ_o'. So every abstract object a satisfying ϕ_o' is such that $a \in ext_{\mathcal{R}}(i)$, by failure of the antecedents. Since i is $\langle -, s \rangle$, no abstract object a is an element of $ext_{\mathcal{R}}(i)$. So every abstract object a which is an element of $ext_{\mathcal{R}}(i)$ satisfies ϕ_o', again by antecedent failure. So, $(a)(a \in ext_{\mathcal{R}}(i) \equiv \phi_o')$. So our procedure for choosing i always guarantees that the conjunction of universal claims is true. Hence our first axiom is true in this interpretation.

Consider, now, $(\exists x)(F)(xF \equiv \phi)$, where ϕ has no free x's. Semantically, we have to show: $(\exists e)(i)(e \in ext_{\mathcal{A}}(i) \equiv \phi') \vee (\exists a)(i)(a \in ext_{\mathcal{A}}(i) \equiv \phi')$. Clearly, the right disjunct is the true one, and this is easily verified. For an arbitrary ϕ', there is an element of $\mathcal{A}$ which has as members all and only the properties i satisfying ϕ'. That's because $\mathcal{A}$ contains all the subsets of $\mathcal{R}$, so pick the one whose elements are just the properties satisfying ϕ' (again, we are appealing to the fact that the following instance of SUBSET is true in the above interpretation: $(\exists y)(z)(z \in y \equiv z \in \mathcal{R} \ \& \ \phi'))$.

It remains, then, to prove the Invariance Lemma:

Proof. By induction on the complexity of ϕ_o.

(1) $\phi_o = Gx$. Since f is fixed, let $f(G)$ be i. $(\rightarrow)$ Assume the antecedent and suppose $f_1(x) = a_1$ and f_1 satisfies ϕ_o, where $f_{1\bar{x}}f$. So, $a_1 \in ext_{\mathcal{R}}(i)$, by the definition of satisfaction. So, $(\exists s)(i = \langle +, s \rangle)$. Now assume $f_2(x) = a_2$ and $f_{2\bar{x}}f$. We need to show $(\exists s)(i = \langle +, s \rangle)$. We already have it. So, $(f')(a)(f'_{1\bar{x}}f \ \& \ f'(x) = a \rightarrow f'$ satisfies $\phi_o)$. $(\leftarrow)$ trivial.

(2) $\phi_o = \sim \psi$. $(\rightarrow)$ Again, assume the antecedent and suppose that $f_1(x) = a_1$ and $f_{1\bar{x}}f$ and f_1 satisfies ϕ_o. So

f_1 fails to satisfy ψ. Now the inductive hypothesis is a biconditional and the existence of f_1 shows that the right side of the biconditional is false. So the left side must be also, that is, $\sim(\exists f')(a)(f'_{\bar{\bar{x}}}f \;\&\; f'(x)=a \;\&\; f'$ satisfies $\psi)$. So, $(f')(a)(f'_{\bar{\bar{x}}}f \;\&\; f'(x)=a \rightarrow f'$ fails to satisfy $\psi)$. Now assume $f_2(x)=a_2$ and $f_{2\bar{\bar{x}}}f$. We want to show that f_2 satisfies ϕ_o. But f_2 fails to satisfy ψ, since it is an f' which satisfies the antecedent of the universal claim. So, f_2 satisfies ϕ_o. $(\leftarrow)$ trivial.

(3) $\phi_o=(\psi\rightarrow\chi)$. Exercise

(4) $\phi_o=(\alpha)\psi$. $(\rightarrow)$ Assume the antecedent and suppose that $f_1(x)=a_1$ and $f_{1\bar{\bar{x}}}f$ and f_1 satisfies ϕ_o. So, every $f'_{1\bar{\bar{\alpha}}}f_1$ is such that f'_1 satisfies ψ. Now assume $g_1(x)=a_2$ and $g_{1\bar{\bar{x}}}f$. We want to show that g_1 satisfies ϕ_o. So, we want to show that every $g'_{1\bar{\bar{\alpha}}}g_1$ is such that g'_1 satisfies ψ. So, let us show, for an arbitrary g'_1, say g^*_1, that g^*_1 satisfies ψ. Pick an f'_1, say f^*_1, such that $f^*_1(\alpha)=g^*_1(\alpha)$. So, f^*_1 satisfies ψ. Hence, $(\exists f'_1)(f'_1(x)=a_1$ and $f'_{1\bar{\bar{x}}}f$ and f'_1 satisfies $\psi)$. We may therefore invoke the inductive hypothesis to get: $(h)(a')(h(x)=a' \;\&\; h_{\bar{\bar{x}}}f \rightarrow h$ satisfies $\psi)$. Since $g^*_{1\bar{\bar{\alpha}}}g_1$, $g^*_1(x)=a_2$. Now if we can show that $g^*_{1\bar{\bar{x}}}f$, we can suppose that g^*_1 is one of the h's and conclude that g^*_1 satisfies ψ. Since $g_{1\bar{\bar{x}}}f$ and $g^*_{1\bar{\bar{\alpha}}}g_1$, it follows that $g^*_{1\,\overline{\overline{x,\alpha}}}f$. But $g^*_1(\alpha)=f^*_1(\alpha)$ and $f^*_{1\bar{\bar{\alpha}}}f$. So $g^*_{1\bar{\bar{x}}}f$. $(\leftarrow)$ exercise. ☒

PART C

I am indebted here again to Dana Scott for suggesting the following interpretation which serves as a model of the modal version of the monadic portion of the theory. The proof runs almost exactly like the consistency proof of the elementary theory, and so we simply sketch the proof below. We shall describe the interpretation in enough detail for the reader to easily fill in the rest. First, we need to describe a few sets and then we will identify the elements of the interpretation in terms of these sets.

Let $\mathscr{W}$ and $\mathscr{E}$ be any two non-empty sets. $\mathscr{W}$ will be the set of possible worlds and $\mathscr{E}$ will be the set we'll use to help construct our possibly existing individuals. The set of possibly existing individuals ("$\mathscr{PE}$") is simply the set of all partial functions from $\mathscr{W}$ into $\mathscr{E}$ (i.e., $\mathscr{PE} = \mathscr{E}^{\cdot\mathscr{W}}$). We let "$\mathscr{p}$" be variables ranging over the elements of $\mathscr{PE}$. These functions have sometimes been known as "individual concepts". A given function takes each possible world and either maps it to an element of $\mathscr{E}$ or is undefined for that world. We use partial functions to represent the real set of possibly existing individuals because the members of this latter set may fail to exist in various worlds. Also, note that in intended interpretations, the partial functions would be constant when defined.

Next, we need to construct our properties. To do this, first consider the set which is the union of two copies of the power set of $\mathscr{PE}$, where the members of the first copy are coded with a plus and the members of the second copy are coded with a minus. That is, first consider the set $\mathscr{K}$:

$$\mathscr{K} = \{+\} \times \mathfrak{P}(\mathscr{E}^{\cdot\mathscr{W}}) \cup \{-\} \times \mathfrak{P}(\mathscr{E}^{\cdot\mathscr{W}}).$$

Now we define $\mathscr{R}$ to be the set of all functions from $\mathscr{W}$ into $\mathscr{K}$, i.e., $\mathscr{R} = \mathscr{K}^{\mathscr{W}}$. We take here the set of all total functions from $\mathscr{W}$ to $\mathscr{K}$ because our logic is two-valued. Now let us define the absolute value of property $\mathscr{r}$ at world $\mathscr{w}$("$|\mathscr{r}|_{\mathscr{w}}$") to be the set $\mathscr{s}$ such that $\mathscr{r}(\mathscr{w}) = \langle +, \mathscr{s} \rangle$ or $\mathscr{r}(\mathscr{w}) = \langle -, \mathscr{s} \rangle$. Here, "$\mathscr{s}$" is a variable ranging over the members of the power set of $\mathscr{PE}$. Finally let $\mathscr{A}$ be the power set of $\mathscr{R}$.

We then define the interpretation $\langle \mathscr{W}, \mathscr{D}, \mathscr{R}, ext_{\mathscr{w}}, ext_{\mathscr{A}}, \mathscr{F} \rangle$ as follows:

$$\mathscr{W} = \mathscr{W}$$
$$\mathscr{D} = \mathscr{PE} \cup \mathscr{A}$$
$$\mathscr{R} = \mathscr{R}.$$

And we define exemplification and encoding as follows:

(1) $\mathscr{o} \in ext_{\mathscr{w}}(\mathscr{r})$ iff either
 (i) $(\exists \mathscr{p})(\mathscr{o} = \mathscr{p}\ \&\ (\exists \mathscr{s})((\mathscr{r}(\mathscr{w}) = \langle +, \mathscr{s} \rangle \vee \mathscr{r}(\mathscr{w}) = \langle -, \mathscr{s} \rangle)\ \&\ \mathscr{p} \in \mathscr{s}))$, or
 (ii) $(\exists \mathscr{a})(\mathscr{o} = \mathscr{a}\ \&\ (\exists \mathscr{s})(\mathscr{r}(\mathscr{w}) = \langle +, \mathscr{s} \rangle))$

(2) $\mathscr{o} \in ext_{\mathscr{A}}(\mathscr{r})$ iff either
 (i) $(\exists \mathscr{p})(\mathscr{o} = \mathscr{p}\ \&\ \mathscr{r} \neq \mathscr{r})$
 (ii) $(\exists \mathscr{a})(\mathscr{o} = \mathscr{a}\ \&\ \mathscr{r} \in \mathscr{a})$.

If we recall here that satisfaction of formulas is defined relative to possible worlds, then these definitions have certain consequences. Let $\mathscr{f}$ be an arbitrary assignment to the variables, and suppose that $\mathscr{f}(F) = \imath$ and that $\mathscr{f}(x) = \mathscr{o}$. Then clause 1 above guarantees that $\mathscr{f}$ satisfies Fx with respect to world $\mathscr{w}$ iff one of the following two conditions holds: (i) $\mathscr{o}$ is some possibly existing object $\mathscr{p}$ and $\mathscr{p}$ is an element of $|\imath|_{\mathscr{w}}$, or (ii) $\mathscr{o}$ is some abstract object $\mathscr{a}$ and the value of $\imath$ at $\mathscr{w}$ is a plus marked set of possibly existing individuals. Since this latter condition is a vacuous one, if some abstract object satisfies an atomic exemplification formula at some world, all abstract objects satisfy the formula at all possible worlds. The Invariance Lemma will guarantee that this happen for all formulas ϕ_0 which have no encoding formulas. Clause 2 has the result that $\mathscr{f}$ satisfies xF with respect to world $\mathscr{w}$ iff one of the following condition holds: (i) $\mathscr{o}$ is some possibly existing object $\mathscr{p}$ and $\imath$ is not self-identical (so this never happens), or (ii) $\mathscr{o}$ is some abstract object $\mathscr{a}$ and $\imath$ is an element of $\mathscr{a}$. Clause (2i) guarantees that the following is true in the interpretation: $\Diamond E!x \rightarrow \Box \sim (\exists F)xF$. Clause 2 in general ensures that the following is true: $(x)(F)(\Diamond xF \rightarrow \Box xF)$. That is because the satisfaction condition for encoding formulas ends up being independent of the possible worlds.

It should be clear that a proof similar to the one constructed in part B can be carried out. We want to show that the following two axioms are true in the above interpretation:

$(\exists F)\Box(x)(Fx \equiv \phi_o)$, where ϕ_o has no free F's and no encoding formulas

$\Box(\exists x)(F)(xF \equiv \phi)$, where ϕ has no free x's.

Semantically, we have to show:

$$(\exists \imath)(\mathscr{w})((\mathscr{p})(\mathscr{p} \in ext_{\mathscr{w}}(\imath) \equiv \phi_o') \ \& \ (\mathscr{a})(\mathscr{a} \in ext_{\mathscr{w}}(\imath) \equiv \phi_o'))$$
$$(\mathscr{w})((\exists \mathscr{p})(\imath)(\mathscr{p} \in ext_{\mathscr{A}}(\imath) \equiv \phi') \vee (\exists \mathscr{a})(\mathscr{a} \in ext_{\mathscr{A}}(\imath) \equiv \phi'))$$

To show that the first is true, we utilize a choice procedure for $\imath$ just like the one constructed in the elementary case. For any given world, the fact that we have got all the (coded) members of the power set of $\mathscr{PE}$ to serve as extensions for our properties makes the left hand conjunct true, while the Invariance Lemma will ensure that our choice of $\imath$ makes the right hand conjunct true. To show that the second axiom is true, just choose $\mathscr{a}$ to be the set of properties $\imath$ which satisfy ϕ' with respect to the given world $\mathscr{w}$. There must be such a set since we've taken the power set of $\mathscr{R}$ to be the set of abstract objects. ⊠ (sketch)

APPENDIX B

MODELLING NOTIONS

Throughout this work, we have talked about notions. Syntactic notions like term, occurrence, the erasure of a formula ϕ, etc., are not all that mysterious – they seem to be BONA FIDE relations among linguistic objects. However, there is a group of metaphysical notions which are rather puzzling. These are all the notions constructed out of the primitive notion of encoding. These notions fall into two groups: defined notions (such as correlation, Form, Monad, World, complete, maximal, etc.) and paradoxical notions (such as exemplifying a property that is not also encoded, exemplifying every property that is encoded, and identity). The former group of notions may not be relations (since the formulas ϕ which would "express' them violate restrictions on λ-formation and RELATIONS) and it is provable that the latter group could not be relations – if they were, some contradiction would be true. The reason these notions are puzzling is because as ontologists, we like to avoid uncategorizable entities ("ontological danglers"). So if these notions are not relations, what are they? Do they have independent ontological status? Or is talk about "notions" just a convenient reification, disguising metalinguistic talk about objects which satisfy definitions?

Even if the answer to the last question is yes, it might be worthwhile to look for a reification procedure whereby we do find some appropriate object in our ontology to code up, or go proxy for our notions. An easy, though risky way to do this would be to add an axiom which asserts that there is a PRIMITIVE encoding relation and a relation behind every one of our DEFINED notions (and leave the paradoxical notions to dangle). Until we have an easy way of confirming the consistency of the theories which result, such a procedure seems suspicious and unsystematic.

There may be a better way, however; one which allows us to find proxies for the primitive notion of encoding, the defined notions AND the contradictory notions. The trick is to build the notional formula in question into the defining formula for an abstract relation. Consider the following two instances of A-OBJECTS:

(1) $(\exists z^{(i,i/p)/p})(F^{((i,i/p)/p)/p})(zF \equiv (\exists x^i)(\exists H^{i/p})(xH \;\&$
$F = [\lambda G^{(i,i/p)/p} GxH]))$

(2) $(\exists z^{(i,i)/p})(F^{((i,i)/p)/p})(zF \equiv (\exists x^i)(\exists y^i)((H^{i/p})(yH \equiv Hx) \;\&$
$F = [\lambda G^{(i,i)/p} Gxy])).$

(1) says there is an abstract $(i, i/p)/p$-relation (between individuals and properties of individuals), z, which encodes a property of such relations, F, iff F is the property of: being a relation which relates an object x^i with a property $H^{i/p}$ it encodes. (2) says that there is an abstract relation among individuals, z, which encodes a property of such relations, F, iff F is the property of: being a relation which relates an object y^i with its correlate x^i. These two abstract relations are unique. It seems reasonable to suppose that they could represent the primitive notion of encoding and the defined notion of correlation, respectively.

Note that by abbreviating a crucial subformula, $(H^{i/p})(yH \equiv Hx)$, in (2), we could rewrite (2) as:

(2)′ $(\exists z)(F)(zF \equiv (\exists x)(\exists y)(Cor(x, y) \;\&\; F = [\lambda G\, Gxy]))$

(3) and (4) are further examples:

(3) $(\exists z^{i/p})(F^{(i/p)/p})(zF \equiv (\exists x^i)(World(x^i) \;\&\; F = [\lambda G^{i/p} Gx]))$

(4) $(\exists z^{i/p})(F^{(i/p)/p})(zF \equiv (\exists x^i)((\exists H^{i/p})(xH \;\&\; {\sim} Hx) \;\&$
$F = [\lambda G^{i/p} Gx])).$

These abstract properties could serve to represent the notions of being a world and being a non-self-correlate, respectively. Necessarily, the latter fails to have a weak correlate – if it did, some contradiction would be true.

It is important to see that this modelling of a contradictory notion doesn't reintroduce Clark's Paradox. Let us use $[\lambda x (\exists H)(xH \;\&\; {\sim} Hx)]$ to denote the abstract property that (4) gives us. That is, we are using what previously had been an ill-formed λ-expression to name a unique abstract object. But we cannot allow this λ-expression to be used in instances of λ-EQUIVALENCE. That is because we have permitted abstract objects of type t to encode abstract properties of type t/p – so we know that there would be an abstract object of type i which encodes $[\lambda x (\exists H)(xH \;\&\; {\sim} Hx)]$ (the abstract property given by (4)). This would be the first move in Clark's Paradox. The second would be to suppose it exemplified $[\lambda x (F)(xF \rightarrow Fx)]$ (an abstract property constructed using an axiom like (4)). But we can stop the paradox from developing to completion by not allowing λ-conversion on the (formerly ill-formed) λ-expressions we have just used to name our abstract properties. Metaphysically speaking, this means that we are not supposing that the abstract properties we have chosen to represent the notions of being a non-self-correlate and being a weak correlate of one's self have in their respective exemplification extensions just the objects which are non-self-correlates or which are weak-self-correlates. And in general, we do not suppose that the abstract relations which represent our defined and contradictory notions have in their exemplification extensions all and only the n-tuples of objects which satisfy the defining formulas of these relational notions. This prevents the paradoxes from being reintroduced.

A question then immediately arises. If the proxy abstract relations do not have the "appropriate" exemplification extensions, why is our modelling procedure worthwhile? Well, the answer is that it is useful. If we generalize our procedure, we can handle data which we couldn't handle before. First, here's our generalization:

> where ϕ is any non-propositional formula, and $\alpha_1, \ldots, \alpha_n$ are any variables of types $t_1, \ldots, t_n$, respectively, then: $[\lambda\alpha_1 \ldots \alpha_n \phi] =_{abbr} (\iota z^{(t_1,\ldots,t_n)/p})(F^{((t_1,\ldots,t_n)/p)/p})(zF \equiv (\exists\alpha_1)\ldots(\exists\alpha_n)(\phi \;\&\; F = [\lambda G^{(t_1,\ldots,t_n)/p} G\alpha_1 \ldots \alpha_n]))$.

This gives us a way to easily name the abstract object which goes proxy for the defined or contradictory notion "expressed" by ϕ. This procedure allow us to represent the following data triad:

A. S believes that Dostoyevsky wrote about the student who killed an old moneylender according to *Crime and Punishment.*

B. S does not believe that Dostoyevsky wrote about Raskolnikov.

C. Raskolnikov is the student who killed the old moneylender according to *Crime and Punishment.*

To get the denotation for the English definite description correct, it must be symbolized as $(\iota x)\Sigma_{CP}(Sx \;\&\; (\exists y)(OMLy \;\&\; Kxy))$. This is the way we did things in Chapter IV, Section 4. But we cannot underline our representing description and turn it into a sense description because it is not constructed out of a propositional formula. "$\underline{(\iota x)\Sigma_{CP}\phi}$" is not a well-defined sense description. So we did not have a definite description in our formal language which had the right denotation and which, when underlined, represented the sense of the English definite description.

But now we can do this. We can suppose that "$\underline{(\imath x)\Sigma_{CP}\phi}$" was defined as follows:

$$\underline{(\imath x)\Sigma_{CP}\phi} =_{df} (\imath z)(F)(zF \equiv F = [\lambda x\ \Sigma_{CP}\phi\ \&\ (y)(\Sigma_{CP}\phi^{y}_{x} \rightarrow y =_E x)]).$$

The λ-expression used in this definition abbreviates a definite description of an abstract property. $\underline{(\imath x)\Sigma_{CP}\phi}$ is the abstract object which encodes just this abstract property. Consequently, we may represent A-B-C as follows:

(A′) $Bsthat\text{-}Wd\underline{(\imath x)\Sigma_{CP}\phi}$ (DICTO)

(B′) $\sim Bsthat\text{-}Wd\underline{r}_s$ (DICTO)

(C′) $r = (\imath x)\Sigma_{CP}\phi$,

This should give the reader a good idea how to handle the data in footnote 12, Chapter VI, Section 1. But what about the following data:

D. S believes that the person who killed an old moneylender according to *Crime and Punishment* was a student.

E. S does not believe that Raskolnikov was a student.

F. Raskolnikov is the person who killed an old moneylender according to *Crime and Punishment.*

To handle data like this, we would need to incorporate the new logical function described in note 11, Chapter VI, Section 1. $\mathscr{ENPLUG}(\imath, o)$ is the proposition that o encodes $\imath$. It would be denoted by "$\tau\rho$", where τ denotes o and ρ denotes $\imath$. We then get:

(D′) $Bsthat\text{-}\underline{(\imath x)\ \Sigma_{CP}(Px\ \&\ (\exists y)(OMLy\ \&\ Kxy))}\,S$

(E′) $\sim Bsthat\text{-}\underline{r}_s S$

(F′) $r = (\imath x)\Sigma_{CP}(Px\ \&\ (\exists y)(OMLy\ \&\ Kxy)).$

But there is still a problem about making this more general. Consider G and H:

G. S believes that Porphyry arrested the student who killed an old moneylender.

H. S does not believe that Porphyry arrested Raskolnikov.

To represent G and H correctly, and in a way which suggests a completely general treatment, we recall that "$\Sigma_{CP}\phi$" abbreviates "$CP[\lambda y \phi]$". So can we represent G and H as:

(G′) $\quad Bs_{that}\text{-}CP[\lambda y\, Ap\underline{(\iota x)\Sigma_{CP}(Sx \;\&\; (\exists u)(OMLu \;\&\; Kxu))]}$

(H′) $\quad \sim Bs_{that}\text{-}CP[\lambda y\, Ap\underline{r}_s]$.

The general solution is to use $\mathcal{ENPLUG}$ on the vacuous property encoded by *Crime and Punishment* to get the proposition that *Crime and Punishment* encodes the vacuous property. This vacuous property is denoted by a λ-expression, $[\lambda y\, Ap(\iota x)\Sigma_{CP}(Sx \;\&\; (\exists u)(OMLu \;\&\; Kxu))]$; but in trying to capture the sense of the English description involved, we have to underline the translating description, thereby denoting an abstract object which encodes an abstract property.

NOTES

INTRODUCTION

[1] Imre Lakatos, [1973], p. 4. I am indebted here to Robert Nola for discussions about Lakatos' work.

[2] In Chapters I–IV, we will use the word "object" to mean "individual". Objects are to be distinguished from relations. However, in Chapters V and VI, the notion of an "object" broadens – individuals and relations (properties, propositions) are all considered to be "objects". Consequently, the abstract objects of Chapters I–IV are just abstract individuals, whereas the abstract objects of Chapters V and VI include abstract individuals and abstract relations.

[3] Strictly speaking, data should not contain technical terms. But these hypotheses of Plato and Leibniz do. So the sense of "data" being used in this case is simply "something to be explained".

[4] We shall not attribute this theory to Russell, though we shall call this view "Russellian" because so many philosophers seem to make the attribution. Russell maintained ([1918]) that the ordinary things we speak of as existing (you, my desk, sub-atomic particles) are "logical fictions" (see p. 253, 270, 271). These things are not "ultimate simples out of which the world is built" (p. 270). He adds facts to the list of things that there are, and he believes that many unreal things like phantoms, hallucinations, and their constituents are in fact real (pp. 257, 274–276). (I am indebted here to Mark Aronszajn for pointing out some of these details.)

[5] See the work of S. Kripke [1963], R. Montague [1974], and M. Cresswell [1973].

[6] See the work of N. Goodman [1951], and H. Field [1980]. We take Quine to be one of the less radical physicalists, but we may not be right in doing so. See Goodman–Quine [1947] and Quine [1948].

[7] See the bibliographical references to the works of Mally, Parsons, Routley, and Castañeda.

[8] Readers who are unfamiliar with abstraction schemata should convince themselves that example (c) does guarantee us an object which exemplifies both properties in question, and no others, despite the fact that a disjunctive condition is involved. A perusal of the following proof should do the trick:

(i)	$(\exists x)(F)(Fx \equiv F=R \vee F=S)$	(NOT′)
(ii)	$(F)(Fa_1 \equiv F=R \vee F=S)$	EE, (i) "a_1" arbitrary
(iii)	Show: $Ra_1 \,\&\, Sa_1 \,\&\, (F)(\sim(F=R \vee F=S) \rightarrow \sim Fa_1)$	
(iv)	Show: Ra_1	
(v)	$Ra_1 \equiv R=R \vee R=S$	UE, (ii)
(vi)	$R=R$	$=$I

(vii)	$R = R \vee R = S$	vI, (vi)
(viii)	Ra_1	$\equiv$E, (v), (vii)
(ix)	Show: Sa_1	
(x)	$Sa_1 \equiv S = R \vee S = S$	UE, (ii)
(xi)	$S = S$	$=$I
(xii)	$R = S \vee S = S$	vI, (xi)
(xiii)	Sa_1	$\equiv$E, (x), (xii)
(xiv)	Show: $(F)(\sim(F = R \vee F = S) \rightarrow \sim Fa_1)$	
(xv)	Show: $\sim(P = R \vee P = S) \rightarrow \sim Pa_1$	"P" arbitrary
(xvi)	$\sim(P = R \vee P = S)$	Assumption
(xvii)	$Pa_1 \equiv P = R \vee P = S$	UE, (ii)
(xviii)	$\sim Pa_1$	$\equiv$E, (xvi), (xvii)

[9] See J. N. Findlay [1933], p. 176. Findlay refers to Meinong [1915], pp. 175-177.

[10] There are a host of good papers by Parsons on the subject: [1974], [1975], [1978], [1979b], and [1979c]. However, the most important statement of his theory is in his book *Nonexistent Objects* [1980].

[11] Parsons [1980], Chapter IV.

[12] Parsons [1980], Chapters VII, VIII.

[13] Findlay describes Mally's theory of determinates in his [1933], pp. 110–112 and pp. 183–184. He cites Mally [1912], pp. 64, 76. Mally's notion of satisfaction is to be understood as our notion of exemplification.

[14] Rapaport [1978], pp. 153–180. For some reason, Rapaport attributes the theory he is working on to Meinong. He calls it a reconstruction of Meinong's theory. His Meinongian objects clearly seem to be Mally's determinations. Although Mally uses the word "determiniert", he also uses "Konstitutiven" ([1912], p. 64). Compare Rapaport's use of "being constituted by".

[15] In what follows, we use the word "abstract" purely as a piece of technical terminology. Also, we take the words "existing", "actual", and "real" to be synonymous.

We could make a long list of entities which, at some time or another, philosophers have supposed to be abstract. It is NOT to be presupposed that the set of abstract objects which we will investigate is (intended to be) identical with the set of objects which some other philosopher pretheoretically intuits to be abstract. Many philosophers have firm intuitions to the effect that certain objects are abstract. However, these intuitions are rarely supported by presenting precise conditions which tell us when there are abstract objects or which tell us when any two abstract objects are identical.

There is both a prerogative and an intellectual obligation to specify how one plans to use the word "abstract". This has been done informally with principles (I) and (II) which follow in the text, and will be done formally in Chapter I (Sections 1, 4). The notion we end up with may not correspond exactly with that of others, but at least it should be clear. And in the course of our investigations, we shall discover that certain objects that other philosophers have taken to be abstract are identifiable among our abstract objects.

[16] For a long time, I thought I had been the first to formulate (I) and (II). But I have subsequently discovered that embedded in Rapaport's dissertation, we find principles roughly similar to these in which Rapaport commits himself to sets. On page 190, we find the

following principle, where "S" ranges over sets of properties, "F" ranges over properties, "o" ranges over Meinongian objects, and "$F\,c\,o$" means that the Meinongian object o is constituted (determined) by F:

(T7a) $(S)(\exists F)(F \in S \rightarrow (\exists o)(F)(F\,c\,o \text{ iff } F \in S))$.

And on page 184, we find:

$$o_1 = o_2 \text{ iff } (F)(F\,c\,o_1 \equiv F\,c\,o_2).$$

CHAPTER I

[1] Every formula is a subformula of itself. If $\phi = (\sim \psi)$, $(\psi\ \&\ \chi)$, or $(\exists \alpha)\psi$, then ψ is a subformula of ϕ. If ψ is a subformula of χ and χ is a subformula of ϕ, then ψ is a subformula of ϕ.
[2] See Alonzo Church [1941].
[3] For convenience, we will read "$E!y$" as "y exists" and "$[\lambda x\, x =_E b]$" as "being E-identical to b", instead of using the more cumbersome readings.
[4] These definitions are all standard.
[5] Compare Parsons [1980], IV, Section 3.
[6] Some of these logical functions can be traced back to Moses Schönfinkel [1924]. Quine developed his predicate operators Der, Inv, inv, Conj, Neg, etc., in his [1960], citing the work of the combinatorial logician H. Curry. I borrowed $\mathcal{PLUG}$ from the 1978 manuscript of Parsons [1980] and used it in the interpretation of the monadic theory of abstract objects developed in my [1978] and [1979a] (Quine had no need for $\mathcal{PLUG}$ since his project was to explain away singular terms). McMichael developed the other logical functions using Quine's operators as prototypes. I learned these algebraic techniques from Alan, and these functions were first used in our [1979b]. Several months later, we discovered that George Bealer was working with logical functions similar to these.

Since [1979b], I have made some minor improvements on these logical functions. $\mathcal{PLUG}$, $\mathcal{UNIV}$, $\mathcal{CONV}$, and $\mathcal{REFL}$ have been turned into families of logical functions by indexing them to the number of the place in the relation on which they operate. This allows us to sharpen up the definitions so that we do not end up generating an infinite number of empty relations like the plugging of a two-place relation in its 300th-place. Also, in Chapter III, $\mathcal{NEC}$ has been constructed and the extensions of all of the functions have been constrained at all possible worlds. Finally, in Chapter V, the functions have been redefined so that they operate throughout the branching type hierarchy.
[7] I am indebted to T. Parsons for pointing this problem out to me. It was not until the second draft that something was finally done about it.
[8] One possibility I have yet to explore is a reference made by Quine in a footnote in [1960]. He says Bernays had developed a system which included axioms. So maybe there is such a theory. Also, Tarski's cylindrical algebra or a polyadic algebra might be relevant here.
[9] I was motivated to construct these definitions after reading Bealer [1981] (in manuscript), Chapter 3. The definition he had constructed to partition his complex terms seemed too complicated. I then realized that by indexing the place numbers to $\mathcal{PLUG}_i$ and $\mathcal{UNIV}_i$ and by ordering the rules for classification, a much simpler procedure for partitioning the λ-expression could be devised. Thanks goes to Alan McMichael for his valuable help in working out many of the details in the following definition.

[10] I'd like to thank M. Jubien for pointing out a flaw in an earlier version of the definition of $\mathscr{I}$-assignment.

[11] See Alfred Tarski [1931] and [1944].

[12] With this definition of satisfaction, we may define what it is for a relation to be **expressible** and define the important concept of **logical consequence** (**model-theoretic**, or **semantic consequence**):

> Relation $\imath^n$ of $\mathscr{I}$ is **expressible** $=_{df} (\exists\phi)$ (ϕ has n free object variables $v_1,\ldots,v_n$ and $(f)(f$ satisfies $\phi \equiv d_{\mathscr{I},f}([\lambda v_1 \ldots v_n \phi]) = \imath^n)$). ψ is a **logical consequence** of $\phi =_{df} (\mathscr{I})(f)$ (f satisfies $\phi \rightarrow f$ satisfies ψ).

[13] I follow Eliot Mendelson [1964], p. 57. Also see Robert Rogers [1971], pp. 87–88.

[14] λ-EQUIVALENCE first appeared with the formal interpretation it presently receives in [1979b] (which was coauthored with Alan McMichael). It replaced the PROPERTIES schema of my [1979a]. Since those early papers I have eliminated primitive identity formulas from the language and drafted a definition of propositional formula. This allows a more elegant formulation of λ-EQUIVALENCE.

[15] The axioms which follow represent the culmination of the process of axiomatization which first began in my [1978] and [1979a]. The axioms found here are basically the same as the ones found in these two early papers, the only difference being that the earlier versions were couched in odd looking languages with primitive identity and which sorted terms denoting (ranging over) A-objects from terms denoting (ranging over) existing objects. The most important axiom, A-OBJECTS, was visualized after reading Parsons, Findlay, and Rapaport.

[16] The reader might wonder here why we have not just defined $x =_E y$ instead of taking $=_E$ as primitive. The reason is as follows: We shall want to be able to form λ-expressions like $[\lambda xy\, x =_E y]$. Had we defined $x =_E y$ as "$E!x$ & $E!y$ & $(F)(Fx \equiv Fy)$", $[\lambda xy\, x =_E y]$ would be ill-formed, due to the presence of the relation quantifier. Recall that we do not allow relation quantifiers into λ-expressions in order to simplify the semantics. The slight loss of elegance which results by having to add a non-logical axiom governing $x =_E y$ is minor compared to the complexity which would result from having to add the technical apparatus required to interpret λ-expressions with relation quantifiers. We make an essential use of λ-expressions in which E-identity appears in Chapter VI, when we model the senses of definite descriptions as objects which encode just the property $[\lambda x \phi\ \&\ (y)(\phi^y_x \rightarrow y =_E x)]$.

[17] In the standard second order predicate calculus, where identity is defined $(x = y =_{df} (F)(Fx \equiv Fy))$, a restricted version of this proper axiom would be a logical theorem. If we were given that for object terms o_1 and o_2, $o_1 = o_2$, we could show that for a formula ϕ with one free object variable v, $\phi^{o_1}_v \rightarrow \phi^{o_2}_v$. Here's how:

Since $o_1 = o_2$, $(F)(Fo_1 \equiv Fo_2)$. In the second order predicate calculus, EVERY formula ϕ with one free object variable can be turned into a property denoting expression $[\lambda v\, \phi]$. Thus we may instantiate the universal F quantifier to get $[\lambda v\, \phi]o_1 \equiv [\lambda v\, \phi]o_2$. But by λ-abstraction, $[\lambda v\, \phi]o_1 \equiv \phi^{o_1}_v$ and $[\lambda v\, \phi]o_2 \equiv \phi^{o_2}_v$. So $\phi^{o_1}_v \rightarrow \phi^{o_2}_v$.

Note that no such proof could be carried out in the object calculus since NOT every formula ϕ with one free variable v can be turned into a property denoting expression $[\lambda v\, \phi]$. Therefore, our identity schema is necessary, since it is not derivable. Our identity schema has even greater significance since it governs identities between relations terms as well. Most treatments of the standard second order predicate calculus fail to discuss adding primitive identity formulas for relation terms, since this would automatically yield the intuitively false consequence that $(F^n)(G^n)(x_1)\ldots(x_n)(F^n x_1 \ldots x_n \equiv G^n x_1 \ldots x_n \rightarrow F^n = G^n)$.

[18] We define:

> a set $\mathscr{S}$ of $\mathscr{I}$-properties is **expressible** $=_{df} (\exists\phi)$ (ϕ has exactly one free F^1-variable and $(f)(f$ satisfies ϕ iff $d_{\mathscr{I},f}(F^1) \in \mathscr{S}$)).

[19] Contrast Parsons [1980], Chapter IV, Section 2; Rapaport [1976], p. 190 T7a; Castañeda [1974], pp. 15–21, C*.1-.7, and *C.1-7; Routley [1979], p. 263.

[20] I have adapted these terms from Rapaport [1978].

[21] We use abbreviations like "*Notion* (x, y)" to remind the reader that these formulas do not abbreviate formulas which can appear behind λ's.

[22] Nor can you: (1) model existing objects as individuals, (2) model A-objects as sets of nuclear properties, (3) model properties as extranuclear properties (where extranuclear properties are conceived as sets of sets of nuclear properties, that is, as sets of A-objects), (4) map down the extranuclear properties in an obvious way so that they become correlated with nuclear, "watered down" versions, and (5) define "x encodes F" as "the nuclear, watered down version of F is an element of x". The reason is that distinct extranuclear properties must sometimes get mapped down to the same, nuclear, watered down version. However, in the theory of abstract objects, if $P \neq Q$, then the object which encodes just P is distinct from the object which encodes just Q. But the above model does not reflect this fact, since these two objects would be identified should the nuclear versions of P and Q be identical.

[23] This result is a fortunate one, for otherwise the theory would be inconsistent. Roughly, the problem would have been as follows. A-OBJECTS effectively yields a one-one function from the power set of the set of properties to the set of abstract objects. Now if we were able to produce a distinct property by plugging up R, for each distinct A-object, we would have a one-one function from the set of abstract objects into a subset of the set of properties. The composition of these two functions would have been in violation of Cantor's theorem, since we would have had a one-one function from the power set of the set of properties into a subset of the set of properties. So our theory rules that the properties produced by plugging are not necessarily distinct.

[24] By "standard", I mean that the verbs are not propositional attitude verbs and that, intuitively, they do not denote higher order properties.

[25] For more on this distinction, see the cited works of Meinong, Mally, Findlay, Parsons, and Routley. Note that we differ with these authors on the property of existence. These authors suppose that it is extranuclear. But for us, A-objects fail to exemplify this property by definition. In Chapter III, it will turn out that they necessarily fail to exemplify this property.

[26] We could add axioms which govern nuclearity, for example, Nuclear(F) & Nuclear$(G) \rightarrow$ Nuclear$([\lambda x\, Fx \,\&\, Gx])$; and, Nuclear$(F) \rightarrow \sim$Nuclear$([\lambda x \sim Fx])$; etc. See Parsons [1979b], pp. 658–660.

CHAPTER II

[1] The material in parts B, C of the Appendix to this chapter is essential for seeing that the axiom we are about to propose is not a logical axiom. Originally, I was misled into thinking

that this was a logical axiom by failing to remember that identity does not work in the object language like it does in the semantics. To see that DESCRIPTIONS is not a logical axiom, consider the Appendix (B, C) and a schematic instance:

$$\psi_v^{(\iota x)\phi} \equiv (\exists !y)\phi_x^y \;\&\; (\exists y)(\phi_x^y \;\&\; \psi_v^y).$$

By the convention employed in Chapter I, Section 4, this abbreviates,

$$\psi_v^{(\iota x)\phi} \equiv (\exists y)(\phi_x^y \;\&\; (z)(\phi_x^z \rightarrow z = y)) \;\&\; (\exists y)(\phi_x^y \;\&\; \psi_v^y).$$

By $\boldsymbol{D}_4$, this abbreviates,

$$\psi_v^{(\iota x)\phi} \equiv (\exists y)(\phi_x^y \;\&\; (z)(\phi_x^z \rightarrow (z =_E y \vee (A!z \;\&\; A!y \;\&\; (F)(zF \equiv yF))))) \;\&\; (\exists y)(\phi_x^y \;\&\; \psi_v^y).$$

Instances of this schema should not be logically true – nothing has been said in our semantics about the primitive relation denoted by "$=_E$". They fail to be valid in the $(\rightarrow)$ direction because of interpretations in which both "$(\iota x)\phi$" denotes an existing object and "$=_E$" denotes an irreflexive relation. They fail to be valid in the $(\leftarrow)$ direction because of interpretations in which there are distinct objects which bear the relation denoted by "$=_E$" to each other but which both satisfy ϕ and ψ.

[2] The results in this section were detailed principally in two early papers [1979c] and [1979e]. I would like to thank Cynthia Freeland for her assistance in locating the relevant passages in Plato's works.

[3] See also *Phaedo*, 100c7-e2, 101a.

[4] An orthodox theorist might suggest that Plato discovered that existential introduction (EI) on predicate terms was a valid rule of inference. This would turn Plato into a language theorist, whereas on our view, he was doing metaphysics.

[5] Contrast Parsons [1980], Chapter VIII, Section 5; also Castañeda [1974].

[6] We are justified in using UE on = I (which is a proper theorem, Chapter I, Section 4) to get $\Phi_P = \Phi_P$ because we can prove that Φ_P has a denotation. In the Appendix to Chapter II, we note that if we can show that some atomic formula containing Φ_P is true, then Φ_P must have a denotation. Theorem 3 provides us with a true atomic formula in which Φ_P occurs. We may therefore appeal to LA4b to instantiate Φ_P into universal claims. LA4b says: $(\alpha)\phi \rightarrow (\psi_\beta^\tau \rightarrow \phi_\alpha^\tau)$, where ψ is any atomic formula and τ contains a description (for details, see the Appendix to this chapter, parts B., C., and D.)

[7] See *Timaeus*, 52c.

[8] See *Timaeus*, 51e, 52a. Also see *The Republic*, 518ff.

[9] See Vlastos' [1954]; and Strang [1971].

[10] Besides using "is" to mean "exemplifies", we also use "is" sometimes to mean "is identical to". And, there will be another defined use of "is", as in "x is a Form of G", "x is a possible world", etc.

[11] For those who prefer to think syntactically, let $\phi = \ulcorner A!x \;\&\; (F)(xF = \chi).\urcorner$ Let $\psi = \ulcorner vG \urcorner$ Then $\psi_v^{(\iota x)\phi} = \ulcorner (\iota x)(A!x \;\&\; (F)(xF \equiv \chi))G.\urcorner$ We then deduced the right side of DESCRIPTIONS using UNIQUENESS and the fact that G satisfies χ.

[12] As usual, with restricted variables:

(i) $(\exists z)\phi_x^z$ abbreviates $(\exists x)(A!x \;\&\; \phi)$

(ii) $(z)\phi_x^z$ abbreviates $(x)(A!x \rightarrow \phi)$.

[13] One suggestion for understanding the ontological status of mathematical objects is to say explicitly which objects exist when formulating the relevant set of axioms. So, for example, we formulate axioms for set theory as follows:

$$\text{NULL } (\exists x)(E!x \,\&\, Sx \,\&\, (y)(y \notin x))$$

$$\text{UNIONS: } (x)(Sx \rightarrow (\exists y)(E!y \,\&\, Sy \,\&\, (w)(w \in y \equiv (\exists u)(u \in x \,\&\, w \in u))))$$

$$\text{POWER: } (x)(Sx \rightarrow (\exists y)(E!y \,\&\, Sy \,\&\, (w)(w \in y \equiv w \subseteq x)))$$

$$\text{SUBSET: } (F)(x)(Sx \rightarrow (\exists y)(E!y \,\&\, Sy \,\&\, (z)(z \in y \equiv z \in x \,\&\, Fz))).$$

It should be clear how to then formulate INFINITE, REGULARITY, and REPLACEMENT. On this kind of formulation, it is provable that there DOES NOT EXIST an object which exemplifies being a set of all non-self-membered sets, though *A*-OBJECTS guarantees that some objects encode this property.

The problem with this suggestion is that it undermines our natural understanding of existence, namely, having a spatio-temporal location.

[14] See Meinong [1904], p. 86 (Section 4 of "Über Gegenstandstheorie").

[15] When we talk about the various senses of an ambiguous property name, we mean the various properties it denotes. We are not referring to its "Fregean" sense.

[16] Necessary beings exist in every possible world or fail to exist in every world – they do not go in and out of existence from world to world. In the next chapter, we redefine *A*-objects as objects which necessarily fail to exist.

[17] Some philosophers may hesitate because they prefer to reserve the term "Platonic existence" to describe properties, relations, and propositions. But we have seen that a certain class of *A*-objects behave like the Forms and this is how we justify calling the kind of existence *A*-objects exemplify "Platonic". Those who now hesitate probably used "Platonic" in connection with properties, etc., in the first place because of the orthodox view that Plato's Forms just are properties.

Those philosophers who still wish to preserve "Platonic existence" for properties, etc., would at least agree that on this usage, the term denotes a (higher order) property of properties. But that would not have bearing on the important question we are now facing – whether it's plausible to think of the negation of the first level property of existence as some SPECIAL kind of existence.

[18] Those philosophers who believe both that properties exist and that sets exist may wonder why we can not dispense with *A*-objects by modelling them as sets of properties. For the reasons why we can not do this, see the discussion at the end of Section 4, Chapter I.

[19] That is, given ϕ^τ_α, we can produce $\sim(\alpha)\sim\phi$ without derived rules as follows: By LA4, we get $(\alpha)\sim\phi \rightarrow \sim\phi^\tau_\alpha$. By the theorem of propositional logic $(\phi \rightarrow \sim\psi) \rightarrow (\psi \rightarrow \sim\phi)$, we get $\phi^\tau_\alpha \rightarrow \sim(\alpha)\sim\phi$.

[20] We also have to modify existential introduction slightly. From ϕ^τ_α, we may immediately infer $(\alpha)\phi$ only if τ contains no descriptions. Otherwise, we first need to know that some atomic formula ψ containing τ is true.

[21] I am indebted to Richard Grandy for noting that the following axiom had to allow for conjunctions of atomic formulas and for suggesting examples of logical truths which would not be derivable using it alone.

[22] Recall note 20.

[23] We have signalled that we are using the semantically primitive notion of identity by switching type styles.

CHAPTER III

[1] Now that we are in a modal theory, we have to face the question of whether our descriptions will vary in denotation from world to world or be "rigid designators". On the first alternative, they would denote at a world w, the unique object satisfying ϕ with respect to w (if there is one). On the second alternative, they would denote at a world w, the unique object satisfying ϕ at w_0 (the base world). We could have two kinds of descriptions in our language – rigid and non-rigid descriptions. However, we shall employ just one type of description, and suppose that all our descriptions are rigid designators.

We do this for two reasons. One is that we will not need non-rigid descriptions in any of our applications. Instead, we shall try to show that rigid descriptions have interesting, heretofore undiscovered, applications. Secondly, by having just rigid descriptions in the language, we can simplify the definition of denotation$_{\mathscr{I},f}$. Since all of the terms of the language will be rigid designators, we need not define the denotation$_{\mathscr{I},f}$ of term τ WITH RESPECT TO world w. Were we to allow descriptions which might change denotations from world to world, we would have to define "$d_{\mathscr{I},f}((\iota x)\phi, w)$". This would force us to revise the entire definition of denotation so that it becomes a binary function.

[2] The technique here is due to Saul Kripke [1963], pp. 83–94.

[3] Note that one can consistently maintain that necessarily true propositions (i.e., propositions $\imath^0$ such that for all worlds w, $ext_w(\imath^0) = T$) need not be identical. For example, the proposition that if Carter is President then Carter is President (i.e. $\mathcal{COND}(\mathcal{PLUG}_1(\mathit{being\ President}, \mathit{Carter}), \mathcal{PLUG}_1(\mathit{being\ President}, \mathit{Carter})))$ need not be identical with the proposition that if Nixon is President then Nixon is President (i.e. $\mathcal{COND}(\mathcal{PLUG}_1(\mathit{beign\ President}, \mathit{Nixon}), \mathcal{PLUG}_1(\mathit{being\ President}, \mathit{Nixon})))$.

Given our statement of the axioms of set theory as in note 13, II, Section 3, we need not believe that there is only one mathematical proposition.

[4] In the definition which follows, and in the definition of denotation, recall that we will often regard propositional formulas ϕ as degenerate λ-expressions $[\lambda v_1 \dots v_n \phi]$ when $n = 0$. That is, $[\lambda\,\phi] =_{df} \phi$. So, in effect, the variables μ, ζ, and ζ range over propositional formulas in so far as they are considered as λ-expressions. For example, according to clause 1, $[\lambda x\, \mathrm{Rab}]$ is classified as the 1st vacuous expansion of $[\lambda\, \mathrm{Rab}]$, which according to the above convention, is just Rab.

[5] Recall that if a definite description fails to denote, the denotation failure is inherited by all the complex terms in which the description appears. Propositional formulas are complex zero-place terms, and hence, they fail to denote propositions if they contain non-denoting descriptions. Note that their satisfaction conditions would still be well-defined.

[6] Had we chosen to interpret our descriptions non-rigidly, this restriction would be unnecessary. However, it is an interesting fact about the logic of rigid descriptions that the rule of necessitation, □I, must be restricted. This prevents the following derivation of a logical theorem which is not true in all interpretations:

(a) $F(\iota x)Gx$ Assumption

(b)	$(\exists y)Gy$	L-DESCRIPTIONS$_1$
(c)	$F(\iota x)Gx \rightarrow (\exists y)Gy$	CP, (a)–(b)
(d)	$\Box(F(\iota x)Gx \rightarrow (\exists y)Gy)$	$\Box$I, (c)
(e)	$\Box F(\iota x)Gx \rightarrow \Box(\exists y)Gy.$	LA7, (d)

(e) is not true in an interpretation in which:

(i) there is a unique object in w_0 exemplifying G,

(ii) this object exemplifies F in every world, and

(iii) there are worlds in which there are no objects which exemplify G.

Unrestricted $\Box$I seems to be the source of the trouble.

I am indebted to Ed Gettier for pointing out this problem; suggestions from Richard Grandy and Max Cresswell have led me to eliminate $\Box$I as a primitive rule and add the modal closures of all the axioms except those governing descriptions. Since we are both allowing open formulas to be assertible and using UI as a primitive rule, I found that you need to add the Barcan formula to successfully derive the restricted version of $\Box$I (see the following footnote).

7 To see that this restricted version of $\Box$I is derivable, first suppose that we have added the modal closures of all of the axioms. Then, the unrestricted rule of $\Box$I could be derived as follows: Suppose $\vdash \phi$. We want to show that $\vdash \Box\phi$. If the proof of ϕ is one line, then ϕ is an axiom. So $\vdash \Box\phi$. If the proof of ϕ is more than one line, then either ϕ was derived from ψ and $\psi \rightarrow \phi$ which appeared on two earlier lines or $\phi = (\alpha)\psi$ and was derived from an earlier line on which ψ appeared. If the former, then the inductive hypothesis is that $\vdash \Box\psi$ and $\vdash \Box(\psi \rightarrow \phi)$ and so by appealing to LA7, $\vdash \Box\phi$. If the latter, then the inductive hypothesis is $\vdash \Box\psi$, so by UI $\vdash (\alpha)\Box\psi$. So by the Barcan formula (LA9), $\vdash \Box(\alpha)\psi$, i.e., $\vdash \Box\phi$.

Since we have unmodalized axioms floating around (the instances of L-DESCRIPTIONS), the base case in the above proof fails. But by adding the restriction to $\Box$I as we have done in the text, we ensure that the base case and the inductive hupothesis never fail. So our restricted version of $\Box$I is a good rule.

8 Here is the proof. Let R be any one-place property. By A-OBJECTS, we have $(\exists x)(A!x \;\&\; (F)(xF \equiv (\exists u)(F = [\lambda y\, Ru] \;\&\; {\sim}uF)))$. Call this object a_5 and suppose ${\sim}a_5[\lambda y Ra_5]$. Then, by definition of a_5, $(u)([\lambda y Ra_5] = [\lambda y Ru] \rightarrow u[\lambda y Ra_5])$. So $a_5[\lambda y Ra_5]$, contrary to hypothesis.

So suppose $a_5[\lambda y\, Ra_5]$. By definition of a_5, for some object, say a_6, $[\lambda y\, Ra_6] = [\lambda y\, Ra_5]$ and ${\sim}a_6[\lambda y\, Ra_5]$. By definition of proposition identity (Section 3, $\boldsymbol{D}_5$), $Ra_5 = Ra_6$. But since a_5 encodes $[\lambda y Ra_5]$ and a_6 does not, $a_5 \neq a_6$.

9 We can distinguish the strong possibility of Socrates' blueprint from the strong possibility of its blueprint as follows:

$$z \textit{ is strongly}_E \textit{ possible} =_{df} \Diamond(\exists x)(E!x \;\&\; Blue(z, x)),$$

Socrates' blueprint is strongly$_E$ possible, whereas its blueprint is not.

10 The material in this chapter was first sketched in my paper [1979d]. The decision to suppose that there were objects which necessarily failed to exist was an agonizing one. It seemed so inelegant at the time, and I tried to construct the objects needed out of the nonexistent objects of the elementary theory. But once I realized that necessarily nonexistent

objects were simply not the kind of objects that could exist, everything seemed to feel a little better. With hindsight, I see that it could not have been any other way.

CHAPTER IV

[1] See F. P. Ramsey, [1927].

[2] The material in this section was first sketched during the writing of the first draft in Fall 1979.

[3] I am indebted to Blake Barley for noting this simplification.

[4] I am also motivated here by the fact that my audience will not consist principally of model-theoretic logicians or mathematicians. The section entitled "Semantics" in Chapter III has been the most technical section so far, and I want readers who have skipped those sections to be able to see that my claims are in fact consequences of the axioms. (As far as metaphysical insights go, they will not have missed much by skipping that section).

[5] The proof is left as an exercise.

[6] See L. Wittgenstein, [1921].

[7] See D. Lewis, [1968].

[8] The material in this section was first developed in my [1979f], written for an independent study on Leibniz. I would like to thank Mark Kulstad for his help in locating certain passages in Leibniz.

[9] The best attempt I know of to make this view precise in orthodox theory is Benson Mates [1968].

[10] Parsons was the first to attempt a precise modelling of monads in an object theoretical framework. See his [1978], [1980]. Parsons' results are proven as metatheorems, with the notion of possible world as primitive. Nevertheless, two of his metatheorems served as the inspiration and prototypes for the results which follow. Castañeda claims to have suggested similar results along these lines in [1974], p. 24. The reader is encouraged to evaluate his suggestion.

[11] Contrast Parsons [1978], p. 147, R_2, and [1980], Chapter VIII, Section 3, Metatheorem 1. Should counterpart theorists reject the analysis of their work presented near the end of Section 2, I have an alternative explanation of why they have come to hold their views. Maybe they have confused strong correlates with their monads. Theorem 8 indicates that it is the monads, not their correlates, which are "world-bound". Such a confusion would put counterpart theorists in good company, for Leibniz may have confused the two as well. See the discussion at the end of this section.

[12] Compare Leibniz [1686b], Section 9, and [1714], Section 56.

[13] Contrast Parsons [1978], p. 147, R_3.

[14] Contrast Parsons [1978], p. 147, R_4, and [1980], VIII, Section 3, Metatheorem 2.

[15] We make no attempt to understand Leibniz's analysis of universal affirmative statements of the form "every A is B". For Leibniz, the quantifier is virtually meaningless. This feature of his logic is not one we wish to preserve.

[16] B. Partee notes that nothing has been said to distinguish stories from essays. The intentions of the author may be relevant here.

[17] Suppose Story (z_0). $(\rightarrow)$ Assume z_0P, for an arbitrary property P. Since z_0P, Vac(P). So for some proposition $Q^0, P=[\lambda y Q^0]$. So $z_0[\lambda y Q^0] \& P=[\lambda y Q^0]$. By EI, $(\exists F^0)(\Sigma_{z_0}F^0 \& P=[\lambda y F^0])$; that is, ϕ_F^P. So by A-DESCRIPTIONS, $(\iota z')(z'F \equiv \phi)P$.

$(\leftarrow)$Assume $(\iota z')(z'F \equiv \phi)P$. Then ϕ_F^P. By reversing the reasoning, z_0P.

Consequently, z_0 and $(\imath z')(F)(z'F \equiv \phi)$ encode exactly the same properties, so they are identical.

[18] Of course, it may be a matter for literary debate as to which propositions are true according to a given story. And the construction of principles which help us to decide the conditions under which a given proposition is true according to a story poses an interesting philosophical problem, one however which is of more pressing concern to a philosopher of fiction than to a metaphysician. The sentences inscribed by the author in the manuscript (or uttered in a storytelling) are not the only sentences which denote propositions true according to the story. By far, the majority of propositions true according to the story are not explicitly stated. Most are the result of an extrapolation process which facilitates communication between the author and his audience. The principles governing the extrapolation process are rather mysterious (see Parsons, [1980], Chapter VII). However, we need not concern ourselves with such mysteries, since the place to begin investigation is with the authorship relation – a relation we take as primitive. To the extent that this relation is unclear, so, too, will our proposal be. However, it should be said that this really reflects a genuine unclarity in our pretheoretical conceptions of the relevant stories.

[19] I believe that this is an important result. It seems to me that much of the potential fiction has for affecting us is bound up in our being able to project ourselves into unreal circumstances which involve objects with which we are already familiar.

[20] *The Seven Per Cent Solution*, a novel by Nicolas Meyer, is supposedly about the secret life of Sherlock Holmes. In this novel, Holmes is a cocaine addict. The novel is meant to be consistent with the Conan Doyle novels.

[21] Contrast Parsons [1980], Chapter III, Section 2.

[22] By *N*-CHARACTERS and *A*-DESCRIPTIONS.

[23] Σ-SUB could be justified as a special instance of a more general principle governing stories which involves the notion of relevant entailment ("$\underset{R}{\Rightarrow}$"). When we add a proposition to our "maximal account" of the story, we should add all the propositions relevantly entailed by this proposition. The following general principle captures this intuition:

$$(\phi \underset{R}{\Rightarrow} \psi \;\&\; \Sigma_s\phi) \rightarrow \Sigma_s\psi.$$

We can derive Σ-SUB from this principle if we suppose, as we surely must, that a consequence of the correct axioms for the predicate logic of relevant entailment will be that $\phi \underset{R}{\Rightarrow} [\lambda x\, \phi^x_o]o$, where o is any object term occurring in ϕ.

[24] D. Lewis poses the following rhetorical question in his [1978], p. 37: "Is there not some perfectly good sense in which Holmes, like Nixon, IS [his emphasis] a real-life person of flesh and blood?" We agree that there is. The sense of "is" in question is "encodes".

[25] If we want to represent "Holmes is a famous fictional detective," we suppose that being famous ("F") is an extranuclear property, and that this is a property Holmes exemplifies. Consequently, we get: Fh & F-detective(h).

However, see Appendix B for a possible method of construing $[\lambda x\, Fx \;\&\; F\text{-detective}(x)]$ as denoting an ABSTRACT property.

CHAPTER V

[1] There may even by a way to model our metaphysical notions like Form, Monad, World, etc., as abstract properties. Recall that we can not guarantee that they are real properties

because their definitions involve encoding formulas. So maybe they are abstract. See Appendix B for the attempt to model them as abstract properties.

[2] Frege [1892], pp. 56–78.

[3] I would like to thank Mark Aronszajn for contributing this example. It was a result of a discussion with Mark that I discovered that the type theory could be used to model the senses of expressions denoting higher order objects. It occurred to me that abstract properties could be the constituents of propositions soon after Mark challenged me to take this type of data more seriously.

[4] See Russell and Whitehead, [1910]. Also Church [1940], pp. 56–58. Our type hierarchy shall bear no resemblance to Church [1951]. In this paper, Church builds a language which, for any given term τ of type α_n, there will be found a term τ', which denotes the sense of τ. Instead of using just two base types (like i and p, as we shall use), Church introduces two infinite lists of base types: $o_0, o_1, o_2, \ldots$ and $\iota_0, \iota_1, \iota_2, \ldots$. The objects in the domain of type o_0 are the truth values and the objects in ι_0 are individuals. Then, where α_n is the type of a given domain of objects, α_{n+1} is the type of the domain of objects which, intuitively, are the "concepts" of the objects of type α_n. For example, the objects of type o_1 are the concepts of truth values (i.e., propositions); objects of type ι_1 are individual concepts; etc. These concepts of type α_{n+1} serve as the senses of terms of type α_n.

In contrast to Church's system, we shall not guarantee that for every term in our formal language, there will be (constructible) a term which denotes its sense. We will concern ourselves only with the senses of terms of natural language. Consequently, we will not need an infinite hierarchy of senses for each given term. We shall get by with just two base types. I am not sure that there is DATA which requires us to suppose that there are senses of terms of a formal language.

[5] I am indebted to Barbara Partee here, whose comments on the syntax of the type theory in [1979h] helped me to be more careful in the final formulation.

[6] The procedure should be clear. Suppose $(\exists\alpha^t)\phi$ is an expression which only violates the second restriction on propositional formulas (i.e. α^t appears as an initial variable somewhere in ϕ). Then let $\psi_1, \ldots, \psi_n$ be the atomic (exemplification) formulas in which α^t is initial. If ψ_i, $1 \leq i \leq n$, is a primitive proposition term F^p, replace ψ_i by TrF^p. If ψ_i is an atomic exemplification formula $\alpha^t\tau_1 \ldots \tau_n$, replace ψ_i by $Ex\alpha^t\tau_1 \ldots \tau_n$. The formula ϕ' which results should be propositional and should "capture the intent" of ϕ.

[7] If we wanted to follow Frege a little more closely, we would define the set of senses of type t, $\mathscr{S}_t$, as follows:

$$\mathscr{S}_t =_{dfn} \{o \mid o \in \mathscr{A}_t \;\&\; (w)((\exists o')(\imath^{t/p})(o \in ext_{\mathscr{A}}(\imath) \rightarrow o' \in ext_w(\imath)) \rightarrow (\exists! o')(\imath^{t/p})(o \in ext_{\mathscr{A}} \rightarrow o' \in ext_w(\imath)))\}.$$

Thus, the objects in $\mathscr{S}_t$ are the abstract objects of type t which have at most one weak correlate. This preserves Frege's intuition that senses DETERMINE at most one object. For example, any A-object of type t which encoded an "individuating" property of type t/p would be in $\mathscr{S}_t$, where

$$\imath^{t/p} \text{ is an individuating property } =_{df}$$

$$(w)((\exists o^t)(o \in ext_w(\imath)) \rightarrow (\exists! o^t)(o \in ext_w(\imath))).$$

Should one decide that Frege's constraint on senses is essential, one would have to redefine sen so that it mapped $\mathscr{D}_i \times \mathscr{N}_t$ into $\mathscr{S}_t$.

[8] Note that if κ^t is not in the vocabulary of o or if o is an abstract individual without

representational capabilities, then we may suppose that $\mathscr{sen}_o(\kappa^t)$ is the null object of type t.
[9] Should it become necessary, we could expand this device by allowing any complex term of type i to serve as subscripts on sense terms. This complication need not be developed for our purposes in Chapter VI.
[10] "z^t" ranges over abstract objects of type t. Consequently, the assignment to "$\underline{(\iota x^t)\phi}$" is the abstract object of type t which encodes just the t/p-property of being the ϕ.
[11] So τ must have the same type as α.
[12] Again, recall "z^{t}" ranges over abstract objects of type t.
[13] I believe that we can do this consistently. However, the theory could be weakened so that A-objects of type t encode only (possibly) existing objects of type t/p. One might think that it is an encouraging sign that the semantics is all set for abstract t/p-properties to have encoding extensions that are non-empty.
[14] I have not yet found an appropriate generalization of the extensional models of the earlier theories which are described in Appendix A.
[15] We sharply distinguish between those English terms which simply lack denotation from those which denote non-existent or abstract objects (see Parsons [1979c]). If there are English proper names which simply lack a denotation, then we need to revise our specification that $\mathscr{F}$ map all the primitive names to a denotation. We will then need to modify LA4a and LA4b accordingly.
[16] The system proposed in this Chapter was first sketched in my [1979h].

CHAPTER VI

[1] D. Kaplan, [1968].
[2] I have borrowed this name, and a few others, from Frege's late essay [1918]. The discussion here does not presuppose familiarity with that work, however.
[3] See K. Donnellan [1974], pp. 3–32, and [1972], p. 377. Also see S. Kripke [1972], p. 302.
[4] Placing such constraints adds the following complexities. As in note 7, Chapter V, Section 2, we would first define the set of senses of type t, $\mathscr{S}_t$, as the abstract objects of type t which have at most one weak correlate. Then we would have to define, for each object o^t in $\mathscr{D}_t$, the set of senses of which o is the unique weak correlate ("$\mathscr{S}_t(o)$"). Then we would require that the $\mathscr{sen}^t_o$ function assign to a given name κ an object drawn from $\mathscr{S}_t(\mathscr{F}(\kappa))$, i.e., an object drawn from the set of senses of which $\mathscr{F}(\kappa)$ is the unique weak correlate. The assigned object would serve as the sense of κ with respect to o'. So where $\mathscr{F}(\sigma) = o'$, $\mathscr{F}(\underline{\kappa_\sigma}) = \mathscr{sen}^t_{o'}(\kappa)$. The object determines $\mathscr{F}(\kappa)$.

Although this succeeds in modelling Frege's ideas, I doubt that language works this way.
[5] We could even imagine a situation in which some other object besides Lauben was the weak correlate of $\underline{\text{Lauben}}_{\text{John}}$.
[6] It follows from the fact that John believes the former and not the latter that these propositions are distinct. This DOES NOT follow from the fact that Lauben $\neq$ $\underline{\text{Lauben}}_{\text{John}}$. Recall the note in Chapter III, Section 4 where we showed that some propositions with distinct constituents might by identical.
[7] Cases in Kripke [1972] and Donnellan [1972], [1974] would be relevant here. However, it is slightly tricky to transpose their arguments designed to refute the Russellian view that names are disguised descriptions into arguments designed to refute the Fregean view that senses determine a unique object as the referent of the term.
[8] Quine [1956].

[9] Note that the English "is the wife of Tully" could be represented either as "$[\lambda x\, Wxt \,\&\, (y)(Wyt \rightarrow y =_E x)]$" or as "$=(\iota x)Wxt$".

[10] That is, any weak correlate of <u>the wife of Tully</u>$_m$ would be the wife of an abstract object. By the AUXILIARY HYPOTHESIS, this never happens.

[11] Mary's belief here is trivial because <u>the wife of Tully</u>$_m$ represents the wife of Tully to Mary. However, you might think that the propositional object of Mary's belief when she believes that the wife of Tully is the wife of Tully is an A PRIORI truth. But consider the A PRIORI truth that <u>the wife of Tully</u> encodes the property of being the wife of Tully. In our formal language, we express this as:

$$\underline{(\iota x)Wxt}[\lambda x\, Wxt \,\&\, (y)(Wyt \rightarrow y =_E x)].$$

Being an encoding formula, this sentence does not denote a proposition. We could, however, develop a new logical function $\mathscr{ENPLUG}$ ("encoding plug"), which maps a property $\mathscr{r}$ and an object $\mathscr{o}$ to the proposition, $\mathscr{ENPLUG}(\mathscr{r}, \mathscr{o})$, which is such that $\mathscr{ext}_{w}(\mathscr{ENPLUG}(\mathscr{r}, \mathscr{o})) = T$ iff $\mathscr{o} \in \mathscr{ext}_{\mathscr{A}}(\mathscr{r})$. This would give us propositions for atomic encoding formulas to denote – propositions which, if true, would be A PRIORI. Then we could represent the object of Mary's belief (above) as an A PRIORI proposition.

We could also represent "Meinong believed that the round square is round" as a relation between Meinong and the proposition $\mathscr{ENPLUG}(\mathscr{being\ round}, \mathscr{the}_{\mathscr{A}}\ \mathscr{round\ square})$. And we could represent "*S* believes that Holmes is clever" as a relation between S and the proposition $\mathscr{ENPLUG}(\mathscr{being\ clever}, \mathscr{Holmes})$ or $\mathscr{ENPLUG}\ (\mathscr{VAC}\ (\mathscr{PLUG}_1\ (\mathscr{being\ clever}, \mathscr{Holmes})), \mathscr{the\ Conan\ Doyle\ novels})$. There are other possibilities here.

[12] There could be a problem here. Is the following triad conclusive data showing that these questions are not independent?

(i) S believes that Dostoyevsky wrote about the student who killed an old moneylender (according to *Crime and Punishment*).

(ii) S does not believe Dostoyevsky wrote about the student who was arrested by Porphyry (in *Crime and Punishment*).

(iii) The student who killed an old moneylender (in *Crime and Punishment*) is the student arrested by Porphyry (in *Crime and Punishment*).

The following representation gets the denotation of the descriptions correct, but does not account for the (apparent?) consistency of the triad:

(i)′ $B\mathit{sthat}\text{-}Wd(\iota x)\Sigma_{CP}(Sx \,\&\, (\exists y)(OMLy \,\&\, Kxy))$

(ii)′ $\sim B\mathit{sthat}\text{-}Wd(\iota x)\Sigma_{CP}(Sx \,\&\, Apx)$

(iii)′ $(\iota x)\Sigma_{CP}(Sx \,\&\, (\exists y)(OMLy \,\&\, Kxy)) = (\iota x)\Sigma_{CP}(Sx \,\&\, Apx)$.

But we can not underline these descriptions to produce sense-descriptions. Should the reader consider this to be a problem, consult Appendix B for a solution.

Also, Parsons has suggested that there is another kind of sentence which causes trouble: "Some biblical prophets are real, some are unreal, and some I am unsure about".

To handle this sentence, let "*B*" denote the Bible, let "*K*" abbreviate the verb "knows", and let "*P*" denote being a prophet. Then consider:

$$(\exists x)(Char(x, B) \,\&\, \Sigma_B Px \,\&\, E!x) \,\&\, (\exists x)(Char(x, B) \,\&\, \Sigma_B Px \,\&\, A!x) \,\&\, (\exists x)(Char(x, B) \,\&\, \Sigma_B Px \,\&\, \sim K\mathit{sthat}\text{-}E!x \,\&\, \sim K\mathit{sthat}\text{-}A!x).$$

[13] Of course, there are other DE DICTO readings:

$$Bj\mathit{that}\text{-}W\underline{w}_j$$
$$Bj\mathit{that}\text{-}\underline{W}_j\underline{w}_j.$$

The same goes for sentence (B) and many of the other sentences which follow. We are now presenting the preferred readings.

[14] There is typical ambiguity here. It should be unobjectionable. See Parsons [1979a].

[15] Some philosophers may prefer to say that Einstein discovered that there were an infinite number of simultaneity relations – one for each frame of reference. So we have to restate our datum sentence as "Einstein discovered that there is no such thing as absolute simultaneity".

[16] "$E!!$" is of type $((p, p)/p)/p$ and "$E!!x$" is defined as: $(\exists y^{(p,p)/p})(E!^{((p,p)/p)/p}y \;\&\; (F^{((p,p)/p)/p})(xF \rightarrow Fy))$. However, it is now necessary to both add "$E!!$" as a simple primitive predicate of type t/p and add the following axiom: $E!!^{t/p}x^t \equiv (\exists y^t)(E!^{t/p}y \;\&\; (F^{t/p})(xF \rightarrow Fy))$. The reason is that as "$E!!x$" is defined, it is not a propositional formula, so "$E!!s$" is not of type p and does not denote a proposition. Consequently, our representation of the datum sentence is ill-formed unless something is done.

[17] In this axiom, "s" is a restricted variable ranging over stories.

[18] Let's suppose that we have restricted our non-logical vocabulary to just the predicates and names used in the story in question. That way, the following principle would not imply that either the sun is shining or it is not the case that the sun is shining is true according to mathematical stories.

[19] I leave it to the reader to determine what the denotation of "117467" is.

BIBLIOGRAPHY

Bealer, George: 1981, *Quality and Concept*, London: Oxford University Press.

Castañeda, Hector-Neri: 1974, 'Thinking and the Structure of the World', *Philosophia* **4**, 3–40.

Church, Alonzo: 1940, 'A Formulation of the Simple Theory of Types', *Journal of Symbolic Logic* **5**, 56–68.

Church, Alonzo: 1941, *Calculi of λ-Conversion*, Princeton: Princeton University Press.

Church, Alonzo: 1951, 'A Formulation of the Logic of Sense and Denotation', *Structure, Method, and Meaning*, New York: The Liberal Arts Press.

Cresswell, Max: 1973, *Logics and Languages*, London: Methuen.

Donnellan, Keith: 1972, 'Proper Names and Identifying Descriptions', *Synthèse* **40**, 356–380.

Donnellan, Keith: 1974, 'Speaking of Nothing', *The Philosophical Review* **83**, 3–32.

Field, Hartry: 1980, *Science Without Numbers: A Defense of Nominalism*, Princeton: Princeton University Press.

Findlay, J. N.: 1933, *Meinong's Theory of Objects and Values*, London: Oxford University Press, (1963).

Frege, Gottlob: 1892, 'On Sense and Reference', translated in Geach and Black (eds.), *Translations of Frege's Philosophical Writings*, Oxford: Blackwell, (1970), 56–78.

Frege, Gottlob: 1918, 'The Thought: A Logical Inquiry', translated A. M. and M. Quinton, *Mind* **65**, (1956), 289–311.

Goodman, Nelson: 1947, 'Steps Toward a Constructive Nominalism', (with W. V. O. Quine), *Journal of Symbolic Logic* **12**, 105–122.

Goodman, Nelson: 1951, *The Structure of Appearance*, Cambridge: Harvard University Press.

Kaplan, David: 1968, 'Quantifying In', *Synthèse* **19**, 178–214.

Kripke, Saul: 1963, 'Semantical Considerations on Modal Logic', *Acta Philosophica Fennica* **16**, 83–94.

Kripke, Saul: 1972, 'Naming and Necessity', *Synthèse* **40**, 253–354.

Lakatos, Imre: 1973, 'Science and Pseudoscience', reprinted in *The Methodology of Scientific Research Programmes*, Cambridge: Cambridge University Press, (1978).

Leibniz, Gottfried Wilhelm: 1679, 'Elements of a Calculus', reprinted in G. H. R. Parkinson (ed.), *Leibniz's Logical Papers*, Oxford: Clarendon Press, (1966).

Leibniz, Gottfried Wilhelm: 1686a, 'General Inquiries about the Analysis of Concepts and of Truths', in *Leibniz's Logical Papers*, ibid.

Leibniz, Gottfried Wilhelm: 1686b, *The Leibniz-Arnaud Correspondence*, H. T. Mason (ed.), Manchester: Manchester University Press, (1967).

Leibniz, Gottfried Wilhelm: 1686c, 'The Discourse on Metaphysics', reprinted in L. Loemaker (ed.), *Leibniz Philosophical Papers and Letters*, Dordrecht: D. Reidel, Holland, (1969), pp. 303–330.

Leibniz, Gottfried Wilhelm: 1714, *The Monadology*, reprinted in *Leibniz Philosophical Writings*, London: J. M. Dent, (1973), pp. 179–195.

Lemmon, E. J.: 1965, *Beginning Logic*, London: Thomas Nelson.

Lewis, David: 1968, 'Counterpart Theory and Quantified Modal Logic', *Journal of Philosophy* **65**, 113–126.

Lewis, David: 1978, 'Truth in Fiction', *American Philosophical Quarterly* **15**, 37–46.

Mally, Ernst: 1912, *Gegenstandstheoretische Grundlagen der Logik und Logistik*, Leipzig: Barth.

Mates, Benson: 1968, 'Leibniz on Possible Worlds', in Rootselaar and Stahl (eds.), *Logic, Methodology and Philosophy of Science*, Amsterdam: North Holland.

Meinong, Alexius: 1904, 'Über Gegenstandtheorie', translated by Levi, Terrell, and Chisholm, *Realism and the Background of Phenomenology*, Glencoe: The Free Press, (1960), pp. 76–117.

Meinong, Alexius: 1915 *Uber Möglichkeit und Wahrscheinlichkeit*, Leipzig: Barth.

Mendelson, Eliot: 1964, *An Introduction to Mathematical Logic*, New York: D. Van Nostrand.

Montague, Richard: 1974, *Formal Philosophy*, New Haven: Yale University Press.

Parsons, Terence: 1974, 'Prolegomenon to Meinongian Semantics', *Journal of Philosophy* **71**, 561–580.

Parsons, Terence: 1975, 'A Meinongian Analysis of Fictional Objects', *Grazer Philosophische Studien* **1**, 73–86.

Parsons, Terence: 1978, 'Nuclear and Extranuclear Properties, Meinong and Leibniz', *Nous* **12**, 127–151.

Parsons, Terence: 1979a, 'Type Theory and Ordinary Language', in S. Davis and M. Mithun (eds.), *Linguistics, Philosophy, and Montague Grammar*, Austin: University of Texas Press.

Parsons, Terence: 1979b, 'The Methodology of Nonexistence', *Journal of Philosophy* **76**, 649–661.

Parsons, Terence: 1979c, 'Referring to Nonexistent Objects', *Theory and Decision* **11**, 95–110.

Parsons, Terence: 1980, *Nonexistent Objects*, New Haven: Yale University Press.

Prior, Arthur N.: 1956, 'Modality and Quantification in S_5', *Journal of Symbolic Logic* **21**, 60–62.

Quine, W. V. O.: 1948, 'On What There Is', *From a Logical Point of View*, Cambrige: Harvard University Press, (1953).

Quine, W. V. O.: 1956, 'Quantifiers and Propositional Attitudes', *The Ways of Paradox*, New York: Random House, (1966), pp. 183–194.

Quine, W. V. O.: 1960, 'Variables Explained Away', *Selected Logical Papers*, New York: Random House, (1966), pp. 227–235.

Ramsey, F. P.: 1927, 'Facts and Propositions', *The Foundations of Mathematics*, New York: Humanities Press, (1950), pp. 138–155.

Rapaport, William: 1976, *Intentionality and the Structure of Existence*, Ann Arbor: University Microfilms, (# 77–1930).

Rapaport, William: 1978, 'Meinongian Theories and a Russellian Paradox', *Nous* **12**, 153–180.

Rogers, Robert: 1971, *Mathematical Logic and Formalized Theories*, Amsterdam: North Holland.

Routley, Richard: 1979, *Exploring Meinong's Jungle and Beyond*, Departmental Monograph #3, Philosophy Department, Research School of Social Sciences, Australian National University, Canberra.

Russell, Bertrand: 1910, *Principia Mathematica* (with A. N. Whitehead), Cambridge: Cambridge University Press.

Russell, Bertrand: 1918, 'The Philosophy of Logical Atomism', reprinted in Robert Marsh (ed.), *Logic and Knowledge*, New York: Macmillan, (1972), pp. 175–282.

Schönfinkel, Moses: 1924, 'On the Building Blocks of Mathematical Logic', in Heijenoort (ed.), *From Frege to Gödel*, Cambridge: Harvard University Press, (1967).

Strang, Colin: 1971, 'Plato and the Third Man', *Plato: A Collection of Critical Essays*, Vlastos (ed.), New York: Doubleday, pp. 184–200.

Tarski, Alfred: 1931, 'The Concept of Truth in Formalized Languages', reprinted in J. H. Woodger (transl.), *Logic, Semantics, and Metamathematics*, Oxford: Clarendon Press, pp. 152–197.

Tarski, Alfred: 1944, 'The Semantic Conception of Truth', *Philosophy and Phenomenological Research* 431–575.

Vlastos, Gregory: 1954, 'The Third Man Argument', *The Philosophical Review* **63**.

Wittgenstein, Ludwig: 1921, *Tractatus Logica-Philosophicus*, London: Routledge and Kegan Paul.

Zalta, Edward N.: 1978, 'An Unorthodox Alternative', unpublished.

— 1979a, 'Alternative Meinongian Semantics', unpublished.

— 1979b, 'An Alternative Theory of Nonexistent Objects', (with Alan McMichael), *Journal of Philosophical Logic* **9**, (1980) 297–313.

— 1979c, 'Plato and Predication', unpublished.

— 1979d, 'A Theory of Objects', unpublished.

— 1979e, 'A Platonic Theory of Forms and an Application to *The Sophist*', unpublished.

— 1979f, 'Monads and Meinong', unpublished.

— 1979g, 'A Semantics for Propositional Attitudes', unpublished.

— 1979h, 'Meinongian Type Theory', unpublished.

INDEX

'Abstract'
  elementary definition 12, 18
  modal definition 60
  typed definition 111
  use of the word 173
Abstraction schemata 8–14, 172
  see also *A*-OBJECTS, λ-EQUIVALENCE, RELATIONS, PROPOSITIONS
Abstracts
  see λ-predicates
'According to the story...' 91ff, 147–148
'Actual' 173
Actual world
  data 2
  definition 81
  semantic 61, 113
A-DESCRIPTIONS (theorem schema) 48
Ambiguity of 'is' 44, 47–49, 83, 90, 94–95, 177
A-object descriptions 48
A-OBJECTS (proper axiom)
  elementary version 34
  modal version 73
  typed version 124
Appearance at worlds 85–86
Argument terms 111
Arrow Elimination (Modus Ponens) 30, 71, 123
Assignments to variables
  discussion 23–24
  elementary definition 26
  modal definition 65
  typed definition 118
Authorship 91ff
AUXILIARY HYPOTHESIS 37–39, 76
Axioms
  Logical 28–30, 68–70, 121–123
  Proper 32–34, 73–74, 124–125

Barcan formula 70, 180
Being 50–52
Blueprints 35 (definiton)
Box Introduction (Rule of Necessitation) 71, 75, 123, 180

Characters of a story 92 (definition)
  Fictional 93
  Native 93
Clark's paradox 158
Cognitive value of a name 108–109
Complete objects 87 (definition)
Complex relations
  of the metaphysics 31–32, 71–72, 123
  of the semantics 20–23, 61–63, 114–116
Compossibility 86 (definition)
Concepts (Leibnizian) 87–90
Conditionalization (primitive logical notion: if-then) 7, 17
  $\mathscr{COND}$ 22, 63, 115
  Conditionalization of ξ and ζ 25–27, 64, 117
Consistency of the theory 158–166
Consistent objects 80 (definition)
Conversion
  $\mathscr{CONV}_{i,j}$ 21, 62, 114
  i, j^{th}-conversion of ξ 25–27, 64, 117
Correlates 35 (definition)
Counterpart theory 83, 181

Dependance (of one formula on another) 30 (definition), 71
De re/de dicto
  data 127–145
  distinction 108
Derivability (see Provability)
Descriptions 40–41, 53–58, 74–75, 97–106, 112–113, 132–135, 137–139
DESCRIPTIONS (proper axiom) 40, 58, 74, 124

Domain of objects 20, 61, 113
Domain of relations 20, 61, 113

E-IDENTITY (proper axiom) 33, 36, 56, 73, 74, 124
E-Identity (see Identity_E)
Embedding 86 (definition)
Encode (primitive metaphysical notion) 12ff
Encoding extension 22, 63, 70, 116
Erasure of a formula 130 (definition)
Exemplification (primitive metaphysical notion) 7ff
Exemplification extension 20, 61, 114
Exist at (a world) 83 (definition)
Existence (primitive theoretical relation) 12, 50–52, 63, 73, 83
Explanation 1–4
Expressible relations 175 (definition)
Expressible sets of properties 176 (definition)

False (under an interpretation) 27, 68
Fictional relations 107, 146
First principles 1, 12–13
Forms 2–3, 41–47, 41 (definition)
Fountain of youth 3, 99

Generalization (see Universal Introduction)
Gold mountain 9, 11, 13, 35, 47–50, 100

Higher order objects 113 (definition)

IDENTITY (proper axiom) 34, 36, 51, 56, 73, 74, 124
Identity 57
  for objects 13, 33, 73
  for properties 13, 29, 69, 121
  for propositions 72, 121
  for relations 29, 69, 121
Identity_E 12, 32–33, 73, 124, 175
Identity Elimination ($=E$) 34
Identity Introduction ($=I$) 33
Impossible objects 145
Impossible relations 145

Individuals 109, 126, 172
Inference rules 30–31, 71, 123
Initial terms 110–111
Interpretations 20ff, 61–64ff, 113–117

λ-EQUIVALENCE 28, 31, 45, 55–58, 69–70, 78, 122, 158–160, 175
λ-IDENTITY 29–30, 69–71, 122
λ-Introduction, Elimination (rules of inference) 31, 71
λ-predicates 16, 18, 23–27, 60–61, 111
L-DESCRIPTIONS (logical axioms) 57, 69–70, 122
Logical axioms
  elementary theory 28–30
  modal theory 68–71
  typed theory 121–123
Logical consequence 175 (definition)
Logical fictions 172
Logical functions 20–22, 61–63, 114–116, 174
Logical theorems 30 (definition), 71
Logical truth 27 (definition), 68

Mathematics 147–153
Mathematical myths 147–150
Mathematical objects 147–153
Mathematical truth 147–148, 152
Maximal objects 78 (definition)
Methodology 1–2
Mirroring (a world) 85 (definition)
Modal logic of encoding 70
Modal schemata 69
Model-theoretic consequence (see logical consequence)
Modus ponens (see Arrow Elimination)
Monads 85 (definition), 84–90

Naive Object Theory 6–10
Names 15, 17, 59, 157
N-CHARACTERS (proper axiom) 93, 95–96, 149–150
Necessary Equivalence 60 (definition), 179
NECESSARY EXISTENCE (proper axiom) 125
Necessary truth 2, 69, 83–84

Necessitation (primitive logical notion) 59, 77
$\mathscr{NEC}$ 63, 116
necessitation of ξ 64, 116, 118
rule of necessitation (see Box Introduction)
Negation (primitive logical notion) 7
$\mathscr{NEG}$ 22, 63, 116
negation of ξ 25–27, 64, 117
NO-CODER (proper axiom) 33, 36, 73–74, 124
Notions 167–171
Nuclear/extranuclear distinction 10–11, 39, 176
Numbers 148–151

Object (primitive metaphysical notion) 7, 12
'Object' 1, 109, 126, 172
Object calculus
elementary 32
modal 73
typed 123–124
Occurrence (of one proposition before another) 92–93
One Over the Many Principle 41–42
Originate (in a story) 93 (definition)

Paradoxes 158–160
Paradox of analysis 143
Participation 42 (definition), 41ff
Partitioning-expressions 24–26, 64–65, 117–118
Platonic Being 43, 52
Platonic Existence 43, 52
Plugging
$\mathscr{PLUG}_i$ 21, 62, 114
i^{th}-plugging of ξ by o 25–27, 65, 118
Possible Objects
possibly existing objects 60 (definition)
strongly possible objects 75 (definition)
weakly possible objects 76 (definition)
Possible worlds 61, 78–84, 79 (definition), 113
Primitive notions 7, 12, 16, 32, 77
Proof 30 (definition), 71
Proof-theoretic consequence (see Provability)
Proper theorems 30 (definition)
Properties 7, 12
Propositional attitudes 4, 107–109, 126–145
Propositional formulas 17–18, 59–61, 110–111, 183
Propositions 61, 72
PROPOSITIONS (logical theorem schema) 72, 123
Provability 30 (definition) 71

Reflection
$\mathscr{REFL}_{i,j}$ 22, 62, 115
i, j^{th}-reflection of ξ 25–27, 65, 118
Relations (primitive metaphysical notion) 7, 12
RELATIONS (logical theorem schema) 31–32, 71–73, 123, 158–160
Representation 108–109, 117
Research programs 1, 5
Rigid designation 103–106, 179
Round square 3, 6, 8, 11, 13, 34, 47–49
Russellian framework 5, 9, 41, 101–102

Satisfaction
elementary definition 27
modal definition 67
typed definition 120
Semantic consequence (see logical consequence)
Senses 4, 108–109, 111–113, 116, 126, 183
Sense descriptions 112–113
Sense terms 111–112, 116, 127
Set theory 151–152, 178
Sophist 44–47
Stories 91 (definition)
STORIES (proper theorem) 91–92, 148
Story Substitution (Σ-SUBSTITUTION) 94
Substitutability 19
Substitution of identicals (see Identity Elimination)

Terms 19, 61, 113
Theorems of Logic (see Logical theorems)
Third Man Argument 43–44
True Belief 130 (definition)

Truth 77 (definition)
Truth at (a world) 81 (definition)
Truth under an interpretation
  elementary definition 27
  modal definition 68
  typed definition 120
Types 109 (definition), 183

Uniqueness 34–35, 40
UNIQUENESS (proper theorem schema) 34–35, 42, 48
Universal Elimination 31
Universal Introduction 30, 71
Universalization primitive logical notion: every) 7, 17
  $\mathscr{UNIV}_i$ 21, 62, 114
  i^{th}-universalization of ξ 25–27, 64, 178

Vacuous expansion
  VAC_i 62, 115
  i^{th}-vacuous expansion of ξ 64, 117
Vacuous properties 77 (defintion)
Validity (see Logical truth)

Weak correlation 76
Well-formed expression 19

GPSR Compliance

The European Union's (EU) General Product Safety Regulation (GPSR) is a set of rules that requires consumer products to be safe and our obligations to ensure this.

If you have any concerns about our products, you can contact us on ProductSafety@springernature.com

In case Publisher is established outside the EU, the EU authorized representative is:

Springer Nature Customer Service Center GmbH
Europaplatz 3
69115 Heidelberg, Germany

Batch number: 10761599

Printed by Printforce, the Netherlands